Environmental Humanities and the Uncanny

Sigmund Freud's essay 'The Uncanny' is celebrating a century since publication. It is arguably his greatest and most fruitful contribution to the study of culture, nature and the environment. *Environmental Humanities and the Uncanny* brings into the open neglected aspects of the uncanny in this famous essay in its centenary year and in the work of those before and after him, such as Friedrich Schelling, Walter Benjamin, E. T. A. Hoffmann and Bram Stoker.

This book does so by focusing on religion, especially at a time and for a world in which some sectors of the monotheisms are in aggressive, and sometimes violent, contention against those of other monotheisms, and even against other sectors within their own monotheism. The chapter on Schelling's uncanny argues that monotheisms come out of polytheism and makes the plea for polytheism central to the whole book. It enables rethinking the relationships between mythology and monotheistic and polytheistic religions in a culturally and politically liberatory and progressive way. Succeeding chapters consider the uncanny cyborg, the uncanny and the fictional, and the uncanny and the Commonwealth, concluding with a chapter on Taoism as a polytheistic religion.

Building on the author's previous work in *Environmental Humanities and Theologies* in bringing together theories of religion and the environment, this book will be of great interest to students and scholars of the environmental humanities, ecocultural studies and religion.

Rod Giblett is Honorary Associate Professor of Environmental Humanities in the School of Communication and Creative Arts, Deakin University. He is the author of many books in the environmental humanities, including *People and Places of Nature and Culture* (2011) and most recently *Environmental Humanities and Theologies* (2018), and is a pioneer in psychoanalytic ecology.

Routledge Explorations in Environmental Studies

Environmental Human Rights
A Political Theory Perspective
Edited by Markku Oksanen, Ashley Dodsworth and Selina O'Doherty

African Philosophy and Environmental Conservation
Jonathan O. Chimakonam

Domestic Environmental Labour
An Eco-feminist Perspective on Making Homes Greener
Carol Farbotko

Stranded Assets and the Environment
Risk, resilience and opportunity
Edited by Ben Caldecott

Society, Environment and Human Security in the Arctic Barents Region
Edited by Kamrul Hossain and Dorothée Cambou

Environmental Performance Auditing in the Public Sector
Enabling Sustainable Development
Awadhesh Prasad

Poetics of the Earth
Natural History and Human History
Augustin Berque

Environmental Humanities and the Uncanny
Ecoculture, Literature and Religion
Rod Giblett

For more information about this series, please visit: www.routledge.com/Routledge-Explorations-in-Environmental-Studies/book-series/REES

Environmental Humanities and the Uncanny

Ecoculture, Literature and Religion

Rod Giblett

Routledge
Taylor & Francis Group
LONDON AND NEW YORK

First published 2019
by Routledge
2 Park Square, Milton Park, Abingdon, Oxon OX14 4RN

and by Routledge
52 Vanderbilt Avenue, New York, NY 10017

Routledge is an imprint of the Taylor & Francis Group, an informa business

British Library Cataloguing-in-Publication Data
A catalogue record for this book is available from the British Library

Library of Congress Cataloging-in-Publication Data
Names: Giblett, Rodney James, author.
Title: Environmental humanities and the uncanny : ecoculture, literature and religion / Rod Giblett.
Description: Abingdon, Oxon ; New York, NY : Routledge, 2019. | Series: Routledge explorations in environmental studies | Includes bibliographical references and index.
Identifiers: LCCN 2018056630 | ISBN 9780367181482 (hbk) | ISBN 9780429059759 (ebk)
Subjects: LCSH: Psychoanalysis and the humanities. | Uncanny, The (Psychoanalysis) | Uncanny, The (Psychoanalysis), in literature. | Ecotheology.
Classification: LCC BF175.4.H85 G53 2019 | DDC 150.19/5—dc23
LC record available at https://lccn.loc.gov/2018056630

ISBN: 978-0-367-18148-2 (hbk)
ISBN: 978-0-429-05975-9 (ebk)

Typeset in Bembo
by Apex CoVantage, LLC

Dedicated to Guan Yin (Goom Yan), the Goddess of Compassion

Dedicated to Chao, Lin, Cong, Kait, the Collins of Cornwall

Contents

Acknowledgements

I am grateful to Elliot Patsoura for sparking my interest in Schelling's uncanny, which I had overlooked in Freud's essay and now comes out into the open (another instance of the uncanny) in the chapter 'The Uncanniness of Schelling's Uncanny.' Elliot also made helpful suggestions about what to read in Schelling's massive body of work (another instance of the monstrous). I am also grateful to Warwick Mules for his critical comments and helpful suggestions that enabled me to clarify the argument, foreground the focus and develop the methodology. Finally, I am grateful to Sandra Giblett for asking 'So what?' about the bibliographic detective story.

Grateful acknowledgement is made to the Board of Fung Loy Kok Institute of Taoism for permission to reproduce material from Fung Loy Kok Institute of Taoism, *A Path of Dual Cultivation: Teachings of the Fung Loy Kok Institute of Taoism,* Toronto: Fung Loy Kok Institute of Taoism, 2008.

1 The uncanniness of Freud's uncanny

The concept/metaphor of the uncanny is arguably the greatest and most fruitful contribution of Sigmund Freud (1856–1939) to the study of culture, nature and the environment. Freud developed the uncanny in his essay of this title first published in 1919 in the psychoanalytic journal *Imago*. It was translated into English by Alix Strachey in 1925 for Freud's *Collected Papers*. This translation was then 'considerably modified' for her husband James' *Standard Edition of the Complete Psychological Works of Sigmund Freud* (Freud, 1955) and republished by Penguin in their 'Penguin Freud Library' 30 years later and so 60 years after the original publication in English (Freud, 1985). A new translation was commissioned and published by Penguin early this century (Freud, 2003).

While a whole swag of Freud's concepts and ideas (such as the Oedipus complex, penis envy and so on) are critiqued and problematized, or pooh-poohed and dismissed as the years go on, the uncanny endures for a century as a useful tool in the toolbox of cultural criticism, literary theory and political and philosophical critique. Freud (1955, p. 219, cf. 2003, p. 124) defines the uncanny as 'that class of the frightening which leads back to what is known of old and long familiar.' 'What is known of old and long familiar' is specifically, though Freud does not mention it, what Randolph (2001b, p. 97) calls 'the mother-infant embrace,' both *in-utero* and *ex-utero*. For Freud (1955, p. 244, 2003, p. 151) the uncanny is literally *unheimlich*, unhomely, but also homely, contradictory feelings which he found associated in the minds of adult males with female genitalia and in his own mind with the first home of individual human life in the mother's womb. The uncanny entails a return of the repressed.

The return of the repressed occurs here and elsewhere for Freud as he does not refer to the mother – his or anybody else's – or to her body, or specifically her womb. The maternal body is repressed in and by Freud in his essay on the uncanny as Randolph (2001a, p. 184, 2001b, p. 97) points out. Freud's uncanny is uncanny in which the repressed maternal body returns. Freud (2003, p. 151) relates that 'it often happens that neurotic men state that to them there is something uncanny about the female genital organs. But what they find uncanny ["unhomely"] is actually the entrance to man's old "home," the place where everyone once lived.' Freud is, of course, referring to the womb, but seems a bit squeamish about doing so explicitly. 'Man's' old home is the womb, whereas the old men's home is (the entrance to) the tomb. The womb is more than a place (where action occurs, or a site where action takes place; it is not a passive receptacle). It is where the processes of life begin and are nurtured. It is also where the first bond occurs *in-utero*, and the mother's body is where the first bond occurs *ex-utero*.

Life *in-utero* is a time and space when, as Freud (1955, p. 235, cf. 2003, p. 143) describes it euphemistically, 'the ego had not yet marked itself off sharply from the external world and from other people.' Freud (1955, p. 235) goes on to remark that 'these factors are partly responsible for the impression of uncanniness.' Both Freud's euphemisms of 'the external world' and 'other people' for the mother's body are factors partly responsible for the impression of uncanniness in Freud's essay on the uncanny. Life *in-utero* and *ex-utero* is a time and space when the ego had not set itself against anything, including the external world, and from other people, such as the mother. The mother herself, as Randolph (2001b, p. 98) puts it following Melanie Klein, is 'the infant's "external world"' and the 'other people' Freud vaguely refers to is the mother. The ego was a part and parcel of the inside world *in-utero* and the uncanny other, the mother, and of the outside world *ex-utero*. In 1919 in 'The Uncanny' Freud misses 'the whole point about mothering' as Randolph (2001a, p. 186) puts it. Twenty years later in his London notes of 1938 Freud (1975, p. 299) gets closer to the point about mothering by mimicking the child: 'The breast is a part of me, I am the breast.' All objects are initially conflated with one another and equated with the mother, or parts of her, especially her breasts, as she/they are the primary object for the young infant. Melanie Klein, as Randolph (2001b, p. 98) puts it, 'gave the mother her due,' as Margaret Mahler did too in her work on psycho-symbiosis and Jessica Benjamin in her work on 'the bonds of love.'[1]

The home place and processes, or perhaps more precisely the 'unhome,' of the uncanny is not only the mother's womb, but also the wetland, the first home of life on the planet in the earth's womb and tomb from both of which new life springs and is nurtured (see Giblett, 1996, chapter 2, 2018b, chapter 1). The wetland is the uncanny place and process *par excellence*. It is also the home, or unhome, of alligators and crocodiles, the 'king' of the tropical wetland, the obverse of the temperate dryland, and the archetypal swamp monsters *par excellence*. In some cultures, the alligator and crocodile are not only monstrous, but also divine. In Judeo-Christian theology, the monstrous and divine are divided, and the monstrous demonized; in the uncanny, the divine and monstrous are re-united.

Chapter 2 traces how monstrous alligators and crocodiles are vehicles and vectors for the uncanny for a variety of writers from the eighteenth to twentieth centuries. The artefacts of colonization, such as the table with the carved alligators (or crocodiles) in the story by L. G. Moberly that Freud recounts in his essay on the uncanny, are one means by which not only the repressed returns, but also the means by which one returns to the repressed. Rather them seeing alligators and crocodiles as monsters to demonize and destroy, chapter two argues for a more environmentally and animal friendly view of them as monstrous figures to fear and respect living in their separate habitats to humans and as companion species on the home planet earth.

The uncanny is applicable not only to the b(l)ack waters of the swamp, but also to the dark underworld of the city which for Freud and Walter Benjamin were an object, or more precisely abject, of horror and fascination, whereas for some nineteenth-century writers on the city its underside was exclusively an object of horror and for some twenty-first century writers the slum is an object of fascination. The dark underside of the city is often figured in terms of colonial places and spaces, such as the jungle and the swamp. Chapter 3 argues that these writers are 'placist' in that they ascribe characteristics to a (human-made) place in the city, such as the slums, that were previously ascribed to a place not made with human hands, or made with ancient human hands (jungle, abyss, nether-land, swamps, etc.), such as darkness, decomposition and disease. As monsters are demonized, so are monstrous places. These places are theologized and moralized as places of evil, the monstrous and the demonic. They need to be detheologized and remythified in much more environmentally friendly terms of, for example, thinking both the city and the swamp as the body.

Other thinkers and writers before and after Freud contributed to the growing body of thinking and writing about the uncanny. In developing his understanding of the uncanny Freud was reliant on those who had gone before, not least the philosopher Friedrich Schelling, who makes a cameo appearance in Freud's essay. Chapter 4 brings into the open the uncanniness of Schelling's uncanny in Freud's essay, including some aspects of the uncanny ignored by Freud (as a lapsed or secular Jew), such as its birth in mythology and polytheism, and its association with the divine and monstrous (ignored by both Freud and Schelling). It also explores the uncanniness of Schelling's uncanny in the work of some Anglophone Schellingians and some writers on Freud, psychoanalysis, the uncanny and the unconscious, such as Vidler's (1992) work on 'the architectural uncanny' of 'the modern unhomely.'

Chapter 4 makes the plea central to *Environmental Humanities and the Uncanny* for valuing polytheism over the monotheisms, and over those monotheistic sectors, contending against each other, often aggressively and violently, in the world today with often disastrous and devastating consequences for people and places with loss of life and ruination of cities and countrysides. It argues for the detheologization and remythification of thinking religion through mythology and for polytheistic poetics as a way of bringing being and living with the earth into closer affinity with nature. It follows in the footsteps and develops the thinking along these lines of Mules (2014). *Environmental Humanities and the Uncanny* prefers polytheology, the dialogic study of multiple, multifarious and polysemic religions, gods and goddesses with a multivocal play of voices traced in performances, to monotheology, the study of one monologic religion and the capital 'G' God with a monopoly on a single, univocal truth inscribed in a written scripture. It also considers monsters and the monstrous (neglected by Schelling and Freud) as and at the intersection of the divine and the human that became separated and the monstrous demonized in the development of the monotheisms, and later rationalism.

The normative modern relationship between monsters and reason is depicted by Francisco Goya in his 1799 etching, 'The Sleep of Reason Produces Monsters.'[2] It depicts a sleeping figure dreaming of monstrous owls, bats and a cat larger than a domesticated one, perhaps a feral cat or a wild feline, such as a leopard. The title and etching imply that monsters are produced in the nightmare of reason when it is sleeping. Monsters do indeed lurk in the sleep of reason. They return to haunt dreams from the monstrous repressed of reason, not because dreams produce the monstrous, but because the sleep of monsters produces reason. Reason tried to put monsters to sleep. Monsters precede reason (as mythology shows). The birth of monsters precedes the birth of reason (as the Bible and *Beowulf* show (see Giblett, 2018b, chapter 2)). The (putting to) sleep of monsters made reason possible. The Christian-induced sleep and repression of monsters produced the bad dream of rationalism and the Enlightenment, in which the repressed monstrous returns (as Castle (1995) shows for that movement and period).[3] The uncanny both returns to the repressed monstrous in the past, and returns the repressed monstrous to presence in the present.

Religion

The uncanny is religion without the boundaries of mere reason. It is contrary to Kant's 'religion within the boundaries of mere reason' of the 1790s (Kant, 1998). This is not to imply that the uncanny is irrational; it is pre-rational and post-rational, before and after reason historically and chronologically, and ex-rational, outside and after reason conceptually and experientially. In the 1940s Georges Bataille invoked Kant for the title of the second part of *Theory of Religion* by entitling it 'Religion within the Limits of Reason' without acknowledging, referencing or discussing Kant (Bataille, 1992, pp. 7 and 63–104). For Bataille as an erstwhile associate of the French surrealists who were aficionados of that which is outside the limits of reason, such as dreams, this seems bizarre.

In the conclusion to *Theory of Religion* Bataille repeated as a throwaway provocation the title of Goya's etching without attribution, comment or elaboration (see Bataille, 1992, p. 113). In the body of this book, Bataille (1992, p. 57) defined religion in Proustian terms as 'the search for lost intimacy.' One wonders, intimacy *with* what? With God, time, place, nature, animals, plants, people, bodies, gods, goddesses? Or all of them? Intimacy, as Bataille implies, does not necessarily have a fixed object or a stable subject, but is, in Julia Kristeva's (1982) terms, abject (between and prior to subject and object). Religion is not the search for a lost object. Religion is the search for lost intimacy *of* and *between* time, place, nature, animals, plants, people, bodies, gods, goddesses and God. The uncanny is the vehicle and vector that enacts this intimacy – whether it is searched for or not. The uncanny can come searching for the dreamer – awake or asleep.

A century before Bataille, Henry David Thoreau (1958, p. 52, 2013, p. 89) defined religion more affirmatively than Bataille when he wrote in a letter dated 8th September 1841 that 'our religion is where our love is.' God is not love (*contra* the New Testament Epistle of 1 *John* 4:8 in which 'God is love'), but religion is love, or more precisely where our love is. When one finds love, one finds religion. When one loses (a) love, as Thoreau did when his beloved older brother died from lockjaw (tetanus) and his sweetheart rejected him on her father's orders, one can become lost, as Thoreau did for a while. The uncanny is the search for lost love, or lost love searching one out, often horrifically, as it did for Thoreau who developed sympathetic symptoms of his brother's lockjaw and who never married. The uncanny is not the search for a lost object of love (as that has gone and is lost), but the search for (lost) love. Thoreau found love (and so religion) in and for the swamps which he loved to walk in and write about, which he entered as the 'Holy of Holies' and where he found 'the strength, the marrow' of the body of the earth (Thoreau, 1982, p. 613, see also Giblett, 1996, pp. 229–239).

Thoreau's swamp religion, however, is not where love is to be found for some. In Western culture swamps and other wetlands, such as marshes, mires, morasses, bogs, lagoons, sloughs, shallow lakes and estuaries, etc., have been seen traditionally, or at least in patriarchal times, as places of darkness, disease and death, horror and the uncanny, melancholy and the monstrous – in short, as black waters. They have often been regarded as home to some sort of horrifying marsh monster or swamp serpent lurking in their murky waters. Up until the 1890s it was thought that the miasmatic vapours that rose from stagnant pools caused malaria (which means literally 'bad air'). The perception persisted from ancient times that miasma also cause melancholia or depression. Wetlands have been filled or drained in order not only to prevent malaria and melancholia, but also because they pose an obstacle to agricultural and urban development.[4] Swamps are where hate and horror are for some, and where melancholy and depression were not for Thoreau (as we will see shortly). The uncanny is the name for this ambivalent love/hate religious relationship with monstrous places, such as swamps or slums (as we will see in chapter 3), and/or with monstrous beings, such as alligators or crocodiles (as we will see in chapter 2).

The uncanny is a religious category, concept and experience, or more precisely it is a religious concept or experience for a culture and time in which God is dead, just like the sublime. As the sublime is a secular theology, so is the uncanny; as the sublime stands in for God as a secular theology (perhaps an oxymoron), so the uncanny stands in for the monstrous divine as a secular demonology; as the sublime is a secular God-substitute, so is the uncanny a secular Satan-substitute; as Satan is a fallen Angel, so the uncanny is the fallen angelic for the world in which God is dead. The sublime and the uncanny can also enact the search for lost intimacy and love in and through technology, or more precisely through sublime communication technologies, such as photography and cinema (as discussed in chapters 5 and 6).[5]

Schelling associated the uncanny with aura, a concept much taken up and employed by Walter Benjamin without making the connection with Schelling. Benjamin discusses aura in relation to the technologies of mechanical reproduction, principally photography and cinema. Chapter 5 brings into the open the uncanniness of Schelling's uncanny in Benjamin's work focusing not only on aura, but also on his concept of the dialectical image, which he sees as operating between 'the what has been' and 'the now of recognizability.' This dialectic is akin to the uncanny as the return of the repressed and the return to the repressed of what J. J. Bachofen called 'the swamp world' presided over by the great mother goddess and that Benjamin discusses in relation to Kafka's writings. The dialectical image is the Marxist or historical materialist analogue and version of the Freudian or psychodynamic uncanny (and vice versa). What Marx called 'the fetishism of commodities' is the industrial capitalist uncanny.

Chapter 5 argues for a remythification of thinking about technology in which the producer and consumer are not enchanted and enthralled by what Benjamin calls 'the phony spell of the commodity,' nor motivated by the drive for mastery over life, death and the earth, but by the desire for mutuality with them. Technology ideally for Benjamin 'is not the mastery of nature but of the relation between nature and human beings.' Benjamin sees this relationship playing out not only in technology, such as photography and the cinema, but also in the city as technology, such as the cities of Naples and Paris in their geologies, geographies and histories. In the case of Paris, a dialectical image occurs between 'the what has been' of Paris as the monstrous marsh in which it was founded and 'the now of recognizability' of Paris as modern metropolis, as what Benjamin called 'the capital of the nineteenth century.' Modern Paris returns to its ancient marshy beginnings and they return as trope in nineteenth-century culture.

The cinema is made for the uncanny. Sitting in the dark in the unhomely home of the cinema watching the interior of unhomely homes, or the exterior of (un)homely places like swamps, with its automatons of disembodied stars displayed on the silver screen on to which are projected desires and fears is certainly uncanny. The repressed returns and the cinema-goer is returned to the repressed irrespective of the content of the film. Some films display and foreground the uncanny more than others in content, style and subject matter. Form and content, structure and material, technology and story coalesce in gothic horror cinema. The uncanny form of film and of the structure of cinema is explored in chapter 5 via the work of Walter Benjamin, especially in relation to the cult of the star and the loss of aura.

Film and cinema are the unhomely home of the uncanny from the beginnings of cinema, such as in the 1922 silent film *Nosferatu*, loosely based on Stoker's *Dracula*; in the *Dracula* movies themselves, beginning famously with Bela Lugosi's portrayal of Dracula (1931); in the original *Mummy* movie with Boris Karloff as 'The Uncanny' (mummy) (1932); and as 'The Monster' in three Frankenstein movies (1931–1938). The uncanny in the movies cries out for a more extensive study than I can give it (and that Beal (2002, pp. 141–192) begins, though the uncanny drops from view and from his vocabulary in his reading of the monster movie, only to re-appear at the end of his instructive study of religion and its monsters).

Freud (2003, p. 135) thought that dolls and automata in mimicking human actions, gestures and behaviour were particularly uncanny as they evoked and elicited both horror and fascination. The boundary and distinction between human and machine are becoming increasingly blurred in the age of communication and information technologies with the cyborg, the cybernetic organism. Chapter 6 considers the cinema-goer as the first cyborg and the cyborg in conjunction with information technologies. The uncanny cyborg is also a persistent topic in visual arts and culture as presented and portrayed in a landmark exhibition in 2001 in Vancouver and resulting catalogue (Grenville, 2001). Human beings connected to and by information technologies are cyborgs, but whether they are (also) symbionts on a symbiotic planet ushering in the symbiocene is another question. This chapter makes the plea for being and doing so.

Freud also thought that fiction was the unhomely home of the uncanny. He also thought that it cried out for a more extensive study than he was able or prepared to give it (Castle (1995) does that for the eighteenth century, while I concentrate on the uncanny in nineteenth- and twentieth-century fiction). Chapter 7 considers the uncanny and the fictional, beginning with the modern adult literary fairy tale. Freud (2003, pp. 135 and 141) suggested that there was 'one writer who was more successful than any other at creating uncanny effects.' He was referring to E. T. A. Hoffmann who he later called 'the unrivalled master of the uncanny in literature.' Freud's proof text is Hoffmann's 'The Sand-Man,' though it is more a proof text for Freud of the castration complex than of the uncanny. In Freud's reading of this story Randolph (2001b, p. 97) argues that 'the insistence on the eyes as repositories of the fear of castration [. . .] is unconvincingly rote, too much like a lecture and not enough like a discovery.'

First published in 1819, a century to the year before Freud's essay on the uncanny, Hoffmann's 'The Mines of Falun' is uncanny (Hoffmann, 1991), not just because the word 'uncanny' appears a couple of times in the English translation by Leonard Kent and Elizabeth Knight, but also as the mines are fascinating and horrifying and smell is the vehicle and vector for evoking and conveying the uncanny. The story theologizes the dark underworld of the mine. In Hoffmann's tale, the protagonist is enchanted by the spell of mineral and metal commodities and is consumed by greed, eventually bringing about his untimely demise.

Similarly, in Bram Stoker's gothic vampire romance *Dracula* first published in 1897 (so more than 20 years before Freud's essay), although the 'u' word only appears three times (Stoker, 2011, pp. 16, 93 and 345), the tale is steeped in the uncanny and the monstrous (as Beal (2002, pp. 124–140) shows in his reading of Stoker's *Dracula* (as we will see in chapter 7)). Along the same lines, the detective story theologizes the dark underworld of the country and city as monstrous, evil places. The heroic role of the detective is to descend into this underworld and bring the penetrating light of reason to bear on its secrets and to bring them to light. This trope is found in the best-selling Sherlock Holmes story *The Hound of the Baskervilles* set predominantly in Great Grimpen Mire and Fergus Hume's even better-selling *The Mystery of a Hansom Cab* set in part in the grimy b(l)ackblocks of late nineteenth-century 'Marvellous Melbourne,' or 'Marvellous Smellbourne.' The dark underworld of the city and the country is often depicted as a slough of despond following in Christian's footsteps in Bunyan's *Pilgrims Progress*.

Despondency is another name for melancholy that is the immanent counterpart to the transcendental sublime and the spiritual counterpart to the psychological uncanny. As the uncanny is a secular theology for a world in which God is dead, so melancholy is a secular spirituality in the lower psychopathological register of the abject. In 1933 Gershom Scholem (1992, p. 81) included a poem in a letter to Walter Benjamin the last line of which concludes that 'where God once stood now stands: Melancholy.' Melancholy is the abject spirituality of a world in which God is dead and, as Terry Eagleton (1986, p. 41) argues, 'the appropriate neurosis for a profit-based society' driven by greed for, and indulging in gluttony of, the earth's resources. Just as the aesthetic and philosophical sublime lifts one up to the heady heights of intellection and theory close to the divine, so the psychological and spiritual melancholy of the uncanny depresses one down into the grotesque lower bodily and earthly strata of slime close to the demonic beneath the black sun of melancholia (see Figure 1, Giblett, 1996, p. 26, and chapters 2 and 7).

By contrast with Scholem, religion for Thoreau is where love is before God stands there and so melancholy cannot stand there. Melancholy for Thoreau does not stand in the swamp, or in nature more generally. On 31st October 1857 Thoreau wrote in his *Journal* (1962, *X*, p. 150) that 'if you are afflicted with melancholy [. . .], go to the swamp.' Thoreau flies in the face of the convention that swamps are melancholic places (see Giblett, 1996, chapter 7). In *Walden* he wrote that 'there can be no very black melancholy to him who lives in the midst of Nature and has his senses still' (Thoreau, 1982, p. 382). Living in the middle of nature (as Thoreau did for some of his life and time) is the prophylactic for melancholy in a world in which God, or one's beloved brother, or love for a sweetheart, is dead, but love lives. There can be no black melancholy to him or her standing in the middle of a swamp and who has his or her senses still to absorb local sounds, smells and sights. On 11th June 1840 Thoreau (1962, *I*, p. 141) wrote in his *Journal* that 'would it not be a luxury to stand up to one's chin in some retired swamp for a whole summer's day.' When one stands in a swamp, melancholy is not there, though the uncanny (the (un)homely) may be).

H. P. Lovecraft, the master of the weird horror story, is also a master of the uncanny in literature (certainly in the twentieth century) to rival Hoffman (in the nineteenth century) as many of his stories show, not just because the word 'uncanny' appears regularly in them (usually once per story), but because he engages with the monstrous and enters that liminal zone between it and the divine that evokes horror and fascination for his characters often via the sense of smell with foetid, fishy or rank odours, but interestingly not in 'The Moon-Bog' (Lovecraft, 2004, pp. 42–49). This story was written in 1921, a mere two years after Freud's essay on the uncanny was first published in German and four years before it was first translated into English. Lovecraft gave Freud's work on the interpretation of dreams short shrift; he may have been more sympathetic to his work on the uncanny. 'The Moon-Bog' relates the story of a wealthy American of Irish background who moves back to Ireland to restore the ruined family castle by the bog. He hated the bog and wanted to drain it. This act of environmental vandalism precipitates the revenge of the bog, or of the monstrous 'thing' of or in the bog, against the American, but not before the bog is evoked in terms of the uncanny.

In J. G. Ballard's second novel, *The Drowned World*, first published in 1962, London has been transformed into a tropical lagoon from increased activity of solar flares melting the ice caps, raising sea levels and flooding low-lying areas in the city and country. In the age of global warming and climate change, *The Drowned World* now seems prescient. It gave birth to a whole series of recent and not so recent dystopian science fiction novels about drowning cities, beginning with George Turner's even more prescient climate change/global warming novel set in Melbourne, *The Sea and Summer*, first published in 1987. None of these progeny, however, have the same reach and depth as *The Drowned World*, especially when it comes to the uncanny and the return of and to the repressed of the waters below and before the city, including its wetlands (see Giblett, 2016).

First published in 1963, the year after J. G. Ballard's *The Drowned World*, Thomas Pynchon's *V* brings together the alligator as a vehicle and vector for the uncanny and the underside of the city as an uncanny place with stories of alligators in the sewers of New York City. These stories date from newspaper accounts of the 1930s and culminate in the late 1950s and early 1960s with Robert Daley's *The World Beneath the City* with a chapter on 'Alligators in the Sewers' (Daley, 1959, pp. 187–189) and ultimately with *V* (Pynchon, 1963). These stories first emerged in the 1930s when Teddy May, the Superintendent of Sewers, or the 'King of the Sewers' as Daley (1959, p. 174) calls him in a chapter devoted to him, 'reigned below ground like a king' (Daley, 1959, p. 175). Pynchon satirizes May in a post-modern parodic reprise of Melville's *Moby Dick* in which the obsessive Captain Ahab hunts the white whale across the greatest ocean of the world so Benny Profane hunts a pinto alligator ('pale white, seaweed black') beneath the greatest city in the world (Pynchon, 1963, p. 111) in an inverted, if not subverted, carnivalesque play on the great American novel.

The uncanny in writing lives on. In the 1980s and 1990s the mantel of the Lovecraftian weird story fell on Thomas Ligotti. Lovecraft is 'a self-admitted early influence on Ligotti work,' according to Jeff Vandermeer (2015, p. ix) in a footnote to his 'Foreword' to the Penguin Classics edition of Ligotti's first two major collections. Indeed, Ligotti's (2015, p. 295) story 'The Last Feast of the Harlequin' is dedicated 'to the memory of H. P. Lovecraft.' Although Ligotti for Jeff Vandermeer (2015, p. x; his emphases) 'came out of the weird and the uncanny genres, he was *passing through* those regions.' To what other regions or destination Ligotti was heading Vandermeer does not specify. It is debatable if the uncanny is a genre as it crosses over and mixes many genres, or types of texts – psychological essay, philosophical treatise, etymological dictionary entry and fictional story. The uncanny is a modality, a way of thinking, seeing and writing, evoked in a variety of genres and discussed in a variety of disciplines and their discourses as these genres suggest.

Perhaps Ligotti was passing through the weird and uncanny to the post-uncanny, or the post-modern uncanny (following the other Thomas, Thomas Pynchon). Like Pynchon in *V*, Ligotti is interested in 'the underbelly of modernity' and 'the blight beneath,' as Vandermeer (2015, p. xi) puts it. Perhaps he was passing through the uncanny to the 'hyper-uncanny' in his synthesis of the banal and the everyday with supernatural horror. Like Lovecraft coining the adverbial 'uncannily' (a word which Ligotti uses once (2015, p. 261)), Ligotti coined the adjectival noun 'hyper-uncanniness' (Ligotti, 2015, p. 49). Ligotti uses 'the uncanny' as noun or as an adjective even more sparingly than Lovecraft.

Vandermeer (2015, pp. ix–x) invokes Kafka, 'the great Bruno Schulz' and other writers as 'iconoclasts' with whom Ligotti is allied (without mentioning Pynchon). Similarly, Chris Mars' cover illustration to the Penguin Classics edition is iconoclastically and uncannily allied to Otto Dix's horrific visions of World War I trench warfare with its dead or wounded soldiers and of its aftermath in decadent Weimar republican Berlin of war cripples and prostitutes. The uncanny evokes the horrifying and fascinating that engages with the corporeal and visceral.

The uncanny persists in recent fiction. In 2017 two 'young adult' urban fantasy horror novels were published with the title of *Uncanny*, one about a wicked witch and wise woman, the other a teenage robot soap opera (Fine, 2017; Gill, 2017). Published also in 2017 was Marjorie Sandor's edited collection *The Uncanny Reader*. It is an anthology of 'stories from the shadows' as announced in its subtitle, beginning predictably with E. T. A. Hoffmann's 'The Sand-Man' (but inexplicably not including L. G. Moberly's 'The Inexplicable' that Freud recounted briefly and described effusively as 'extraordinarily uncanny').

Pursuing or being pursued by monstrous animals and killing them or being killed by them is the stuff in trade of the uncanny in literature, legend and culture. The legend of Saint George killing the evil dragon is a well-known case in point. It's in books and statues and on coins. Throughout the Commonwealth, institutions and streets are named 'St Georges.' St George is the patron saint of England, the Church of England and the Scouts. He is a monumental figure of monotheistic Christianity while the dragon is a monstrous creature of polytheism who inhabits wetlands, both of which need to be destroyed to establish the Christian religion in the Christian nation and the Christian colony in the Christian empire. The legend is a foundation myth of English nation-building and British colonization and empire-building, including the colonization of nature in general, and wetlands in particular.

Chapter 8 calls for the decolonization of dragons as evil monsters and their remythification as divine monsters.[6] It begins with a brief recount of the legend and goes on to undertake its decolonization by pointing to its retelling it from an environmentally and animal friendly point of view (Giblett, 2018a, 2018c). It then draws on the theory of the history, politics and role of the Commonwealth as outlined by Cultural Studies pioneer Stuart Hall (1987), by the now largely forgotten Patrick Gordon Walker (1962) and recently by Philip Murphy (2018), none of whom consider Christianity as central to the projects of colonization, especially of nature and places such as wetlands, and decolonization, including nature and wetlands.

The Commonwealth of Christianity with a shared language and a common tradition and institution in the Christian churches is the handmaid to the old British political and economic empire with their corporations currently colonizing the world, the planet and beyond in the Commonwealth with the other three evil empires (China, Russia and the United States and associated corporations). Colonization is as much about the colonization of nature as it is about the colonization of 'the natives,' and the colonization of nature is just as much about the colonization of 'swamps' as the colonization of 'the bush.' It is also just as much about the triumph of monotheism over polytheism, of colonizing monotheistic Christianity over polytheistic local religions of the colonized. The Commonwealth is also a Commonwealth of counter-culture of writers and other cultural workers who work in solidarity with each other against the four evil empires and for the decolonization of people, places and the planet. In this respect 'post-colonial' cultural and literary studies, including post-colonial eco-criticism, seem premature as decolonization is not complete, is ongoing and may never finish – just as colonization is ongoing. Colonization can be defined as the appropriation and exploitation of people, places and the planet for the profit and pleasure of the more powerful few at the expense of the poverty and pain of the less powerful many.

Environmental Humanities and the Uncanny brings into the open neglected aspects of the uncanny in Freud's famous essay and in the work of those before and after him, such as Friedrich Schelling, Walter Benjamin, E. T. A. Hoffmann and Bram Stoker. It does not go over the same dry, old ground covered in their studies of the uncanny by Cixous (1976), Royle (2003) and Masschelein (2011). It goes into different new, wetland by concentrating on aspects of the uncanny and reading stories of it neglected or overlooked by them, such as wetlands as the uncanny place *par excellence*, alligators and crocodiles in Freud's recount of L. G. Moberly's 1917 story about a table with carved crocodiles on it (Moberly, 1917, 1991 discussed in chapter 2); Freud's anecdote about walking in the red-light district of an Italian town (considered in chapter 3); and Schelling's uncanny, especially his discussion of mythology and polytheism (considered in chapter 4).

Detheologizing and remythifying religion means moving away from monotheology to polytheology, the study of many religions. One way of doing so is via Taoism, the way of the way (or Tao). Taoism is polytheistic. It has many gods and goddesses. Polytheism is alive and well in the Fung Loy Kok Institute of Taoism/Taoist Tai Chi Society, a not-for-profit organization active in over 20 countries. Chapter 9 concludes *Environmental Humanities and the Uncanny* by presenting and discussing the living polytheism of this organization dedicated to bringing the beliefs, practices and benefits of the 'three religions of China' (Buddhism, Confucianism and Taoism) to all communities. One of the aims and objectives of the Society is to promote cultural exchange by bringing the riches of Chinese culture to other cultures. Founded in Canada in 1970 by Master Moy Lin-shin, the Society is the largest volunteer tai chi organization in the world. Chapter 9 draws on the publications of the Society, including transcripts of talks by Master Moy and translations of Chinese texts, to present these beliefs and the benefits of practising the Taoist arts of health, including tai chi and meditation. Information about its background, classes and other activities is available on its website: www.taoist.org. I am a member. New members are welcome.

Notes

1 For further discussion of the work of Melanie Klein, Margaret Mahler and Jessica Benjamin see Giblett (2011, chapters 9 and 12).
2 Reproduced in Castle (1995, figure 1.5, p. 18) with the title 'The Nightmare of Reason' without comment or analysis. See Beal (2002, p. 119) for a discussion of Goya's etching.
3 For Christianity, monsters and dragons, see Giblett (2018b, chapter 2, 'Theology of Wetlands and Marsh Monsters' and chapter 3, 'Theology of Dragons and Monstrous Serpents').
4 Each of these aspects of wetlands receives a systematic, often chapter-by-chapter, treatment in Giblett (1996).
5 For the sublime and the uncanny, see Giblett (1996, chapter 2, 'Philosophy in the Wetlands: The S(ub)lime and the Uncanny'). For sublime communication technologies, see Giblett (2008, especially pp. viii–x).
6 For environmentally and animal friendly retellings of the legend of St George and the Dragon and of the story of Beowulf and the Dragon, see Giblett (2018a and c).

References

Ballard, J. 1983. *The Drowned World.* London: Dent.

Bataille, G. 1992. *Theory of Religion.* R. Hurley, trans. New York: Zone.

Beal, T. 2002. *Religion and Its Monsters.* New York: Routledge.

Benjamin, W. and G. Scholem. 1992. *The Correspondence of Walter Benjamin and Gershom Scholem 1932–1940.* G. Scholem, ed. G. Smith, trans. Cambridge, MA: Harvard University Press.

Castle, T. 1995. *The Female Thermometer: Eighteenth-Century Culture and the Invention of the Uncanny.* New York: Oxford University Press.

Cixous, H. 1976. Fiction and Its Phantoms: A Reading of Freud's *Das Unheimliche* (The 'Uncanny'). *New Literary History*, 7(3), pp. 525–548.

Daley, R. 1959. *The World Beneath the City.* Philadelphia: J. B Lippincott.

Eagleton, T. 1986. *William Shakespeare.* Oxford: Basil Blackwell.

Fine, S. 2017. *Uncanny.* New York: Skyscape.

Freud, S. 1955. The 'Uncanny.' In: *The Standard Edition of the Complete Psychological Works of Sigmund Freud, Volume XVII (1917–1919): An Infantile Neurosis and Other Works*, A. and J. Strachey, trans. London: Hogarth, pp. 217–256. First published 1919.

Freud, S. 1975. Findings, Ideas, Problems. In: *The Standard Edition of the Complete Psychological Works of Sigmund Freud, Volume XXIII.* J. Strachey, trans. London: Hogarth, pp. 299–300. First published 1938.

Freud, S. 1985. The 'Uncanny.' In: *Art and Literature*. A. and J. Strachey, trans. Harmondsworth: Penguin, pp. 335–376. First published 1919.

Freud, S. 2003. *The Uncanny*. D. Mclintock, trans. London: Penguin. First published 1919.

Giblett, R. 1996. *Postmodern Wetlands: Culture, History, Ecology*. Edinburgh: Edinburgh University Press.

Giblett, R. 2008. *Sublime Communication Technologies*. Basingstoke: Palgrave Macmillan.

Giblett, R. 2011. *People and Places of Nature and Culture*. Bristol: Intellect Books.

Giblett, R. 2015. *Canadian Wetlands: Places and People*. Bristol: Intellect Books.

Giblett, R. 2016. *Cities and Wetlands: The Return of the Repressed in Nature and Culture*. London: Bloomsbury.

Giblett, R. 2018a. *The Dragon and Saint George: A Fairy Tale Novella*. Cambridge: Vanguard Press.

Giblett, R. 2018b. *Environmental Humanities and Theologies: Ecoculture, Literature and the Bible*. London: Routledge.

Giblett, R. 2018c. *Tales of Two Dragons*. London: Austin Macauley.

Gill, D. 2017. *Uncanny*. New York: Greenwillow Books.

Grenville, B., ed. 2001. *The Uncanny: Experiments in Cyborg Culture*. Vancouver: Arsenal Pulp Press.

Hall, S. 1987. *Pictures of Everyday Life: People, Places and Cultures of the Commonwealth*. London: Routledge.

Hoffmann, E. 1991. The Mines of Falun. In: *Spells of Enchantment: The Wondrous Fairy Tales of Western Culture*. J. Zipes, ed., L. Kent and E. Knight, trans. New York: Viking Penguin, pp. 304–324. First published 1819.

Kant, I. 1998. *Religion Within the Boundaries of Mere Reason and Other Writings*. A. Wood and G. di Giovanni, trans. and ed. Cambridge: Cambridge University Press. First published in 1793.

Kristeva, J. 1982. *Powers of Horror: An Essay on Abjection*. L. Roudiez, trans. New York: Columbia University Press.

Ligotti, T. 2015. *Songs of a Dead Dreamer and Grimscribe*. New York: Penguin.

Lovecraft, H. 2004. The Moon-Bog. In: *The Dreams in the Witch House and Other Weird Stories*. S. Joshi, ed. New York: Penguin, pp. 42–49. First published in 1921.

Masschelein, A. 2011. *The Unconcept: The Freudian Uncanny in Late-Twentieth-Century Theory*. Albany, NY: State University of New York Press.

Moberly, L. G. 1917. Inexplicable. *The Strand Magazine*, 54(324), pp. 572–581.

Moberly, L. G. 1991. Inexplicable. In: *Strange Tales from the Strand*. J. Adrian, ed. Oxford: Oxford University Press, pp. 183–195. First published 1917.

Mules, W. 2014. *With Nature: Nature Philosophy as Poetics Through Schelling, Heidegger, Benjamin and Nancy*. Bristol: Intellect Books.

Murphy, P. 2018. *The Empire's New Clothes: The Myth of the Commonwealth*. London: C Hurst.

Pynchon, T. 1963. *V.* New York: Random House.

Randolph, J. 2001a. Looking Back at Cyborgs. In: *The Uncanny: Experiments in Cyborg Culture*. B. Grenville, ed. Vancouver: Arsenal Pulp Press, pp. 182–186.

Randolph, J. 2001b. Transgressed Boundaries: Potent Fusions and Dangerous Possibilities. In: *The Uncanny: Experiments in Cyborg Culture*. B. Grenville, ed. Vancouver: Arsenal Pulp Press, pp. 95–99.

Royle, N. 2003. *The Uncanny*. New York: Routledge.

Sandor, M., ed. 2017. *The Uncanny Reader: Stories from the Shadows*. New York: St Martin's Griffin.

Scholem, G. 2012. *On Jews and Judaism in Crisis: Selected Essays*. W. Dannhauser, trans. Philadelphia: Paul Dry.

Stoker, B. 2011. *Dracula*. Oxford: Oxford University Press. First published 1897.

Thoreau, H. 1958. *The Correspondence of Henry David Thoreau*. W. Harding and C. Bode, eds. Washington Square, NY: New York University Press.

Thoreau, H. 1962. *The Journal of Henry D. Thoreau, Volumes I- XIV*. B. Torrey and F. Allen, eds. New York: Dover.

Thoreau, H. 1982. *The Portable Thoreau*. C. Bode, ed. New York: Viking.

Thoreau, H. 2013. *The Correspondence of Henry D. Thoreau: Volume 1: 1834–1848*. R. Hudspeth, ed. Princeton: Princeton University Press.

Turner, G. 2013. *The Sea and Summer*. London: Victor Gollancz. First published 1987.

Vandermeer, J. 2015. Foreword. In: *Songs of a Dead Dreamer and Grimscribe*. New York: Penguin, pp. ix–xiv.

Vidler, A. 1992. *The Architectural Uncanny: Essays in the Modern Unhomely*. Cambridge, MA: MIT Press.

Walker, P. 1962. *The Commonwealth*. London: Secker & Warburg.

2 Alligators, crocodiles and the monstrous uncanny

Encounters between humans and crocodiles unfortunately occur in Australia with monotonous regularity and often with disastrous consequences for both parties, but also with cultural implications for how humans and animals co-habit planet earth. This is especially the case for the way in which crocodiles are portrayed, including in the Bible and usually as some sort of monster. For instance, in north Queensland in October 2004 a man was dragged out of a tent by a crocodile only to be saved from a worse fate by the valiant efforts of a grandmother who jumped on the back of the croc who released the man and then proceeded to attack her. The print media headlined the story and captioned the accompanying photo with 'Gran who beat off croc attack' (2004). This headline gave a curiously Australian, and horrifyingly real, inflection to the immortal lines of a Tony Joe White song, 'Polk Salad Annie/'gator's got your Grannie/ chomp, chomp' (White, 1997). The newspaper story described the crocodile as 'a bloodthirsty predator.' This description of the crocodile was only jumping on the same bandwagon as its cousins in television news as they had already referred to the crocodile earlier in the week as 'the four-metre monster.' In both constructions of the crocodile as bloodthirsty predator and monster not only its size, but also its use of its jaws and teeth as a potentially lethal weapon and the fearful possibility of being eaten were placed on the menu for the delectation of the media consumer over breakfast or dinner who could savour with relief that they were safe from being eaten.

This event and its media aftermath hark back to other incidents and representations involving crocodiles, such as Baby Bob and Steve Irwin, especially as there was a baby involved in the most recent incident. Fears for the safety of Baby Bob and the vilification of Steve Irwin from Indianapolis to Indooroopilly highlight human's fascination with, and fear of, crocodiles and their cousins, alligators. They also highlight the visceral nature of human bodily being and experience, particularly when it comes to food and feeding. With Steve holding a chicken in one hand to feed the croc and Bob in the other *not* to feed it, but with the possibility that it could feed on him, the boundary between being food and not food was evident for all to see, especially as baby and chicken were about the same size and colour. This event raised fears of an oral kind, about who gets to eat and who gets to be eaten.

It also raised again the horrifying possibility that a croc might take a baby, just as a dingo took one, too, as with Azaria Chamberlain in 1980. And just as she was taken from a tent, so in the most recent case the baby was in the tent. As in the Lindy Chamberlain case, the Steve Irwin case also brought about the equally terrifying actuality of another moral panic and trial by media with Steve accused of careless parenting like Michael Jackson, or attempted son-icide with the croc as weapon, or accomplice, or both. The portrayal of the croc as some sort of orally sadistic monster made it well-suited, if not typecast, to play this role which it has been playing for a long time – but only since humans have appeared on the earth long after it.

In this chapter I argue that the typecasting of the alligator and the crocodile as orally sadistic monsters is a projection of human desires and fears on to these non-human beings. These desires and fears of an oral nature are tied up with what Freud calls 'the uncanny.' The concept/metaphor of the uncanny is arguably Freud's greatest and most fruitful contribution to the study of culture, and to what I call, following Alexander Wilson, the cultures of natures (see Giblett, 2011, chapter 1). Yet the uncanny is interestingly and symptomatically absent from the classic dictionary of psychoanalytic terms (Laplanche and Pontalis, 1973), and from a relatively recent account of the development and context of Freud's thought (Makari, 2008).

The most recent intellectual and cultural biography of Freud devotes only half a page out of its 500 pages and a footnote in a chapter on 'Lumière sombres' ('Dark Enlightenment') to a discussion of the uncanny with the German *unheimlich* translated into French as 'l'inquiétante étrangeté' and as the 'l'inquiétante familier' following previous translations of Freud's essay into French (Roudinesco, 2014, pp. 269, 269 n.2 and 544). The French terms in cognate English terms of 'the strange disquietude' and 'the disquietingly familiar' are translated as 'the uncanny' in Roudinesco (2016, pp. 215, 468 n.2 and 548).

The uncanny fares even worse in Joel Whitebook's recent intellectual biography of Freud in which it is not discussed and Freud's essay on it is not even listed in the bibliography (Whitebook, 2017). The uncanny tends to be repressed in the secondary psychoanalytic literature. By contrast, the uncanny receives a full-length book study in cultural and literary studies (see Royle, 2003). It is also invoked in ecocultural studies, especially for the wetland as the uncanny place *par excellence* (see Giblett, 1996, especially chapter 2) and for the underside of the city (as we will see in the following chapter).

For Freud, the alligator and the crocodile portrayed as orally sadistic monsters are vehicles and vectors of the uncanny. This combination gives rise to what could be called 'the monstrous uncanny' in which the fascinating and horrific are projected onto, and embodied in, an orally sadistic monster. The uncanny, I have argued elsewhere, counters the aesthetics of the sublime, the picturesque (pleasing prospects) and the beautiful (see Giblett, 1996, chapter 2 and 2011, chapter 3). Whereas the latter three privilege the distancing sense of sight, the uncanny engages the sense of smell which is much more up close and personal, though not as immediate as touch and taste. Smell is often commented upon in encounters, real or imagined, with an alligator or crocodile.

The monstrous uncanny, however, not only engages the olfactory but also the oral and tactile. The uncanny associated with smell (as Freud did) is a sublimation to some extent of the uncanny associated with taste and touch. The uncanny is evoked by what is not seen, which could not only be what is smelt, but also what is tasted and touched, both of which are involved when being eaten. The fear of being killed and eaten, and the desire not to be, are understandable, and certainly characterize the human side of the relationship with alligators and crocodiles. It also presumably characterizes the other side too. The relationship between eater and eaten is ultimately non-reciprocal: only one being gets to eat and the other to be eaten. When humans venture into the habitat of alligator or crocodile they can be prey and 'being prey' as Val Plumwood describes it is a terrifying experience, but also instructive as it was for her (as we shall see). Being prey highlights the non-reciprocal nature of the relationship. Besides consuming alligators and crocodiles as meat, humans also consume their habitat by destroying wetlands in a colonizing and non-reciprocal relationship.

The monstrous uncanny is also associated with the colonial unconscious, whether it be with William Bartram's and John Muir's encounters with an alligator in a Florida swamp, or Val Plumwood's and Sigmund Freud's accounts of stories about crocodiles in a New Guinea swamp. The return to the repressed is a return not only to the individual's own repressed, but also to the culture's repressed. Both of these are figured in all these stories in association with the alligator or crocodile as an orally sadistic monster and the swamp as a grotesque place. Rather than reproducing this figuration of monstrosity, this chapter concludes by arguing for a relationship with alligators and crocodiles characterized by mutuality in which they and their habitat are respected and conserved. Animals, as Whatmore (2002, p. 32) advises, are 'best considered as strange persons, rather than familiar or exotic things.' In other words, they are best considered as what Haraway (2008) calls companion species, or as uncanny beings, rather than as monsters.

Freud and the smelly uncanny

Alligators and crocodiles have been living on the earth for 200 million years, much longer than most other currently surviving, similar-sized genus of the animal kingdom. They are truly a 'blast from the past.' As such, they are vehicles and vectors for the uncanny. Freud (1985, p. 340) defined the uncanny as 'that class of the frightening which leads back to what is known of old and long familiar.' The uncanny is not only a return to the past, but also in quasi-Freudian terms 'a return *to* the repressed,' including the colonial repressed (as I have argued elsewhere; see Giblett, 1996). What was repressed for the nineteenth-century, petit-bourgeois and Viennese Freud was invariably sexual in nature.

More generally in patriarchal culture the repressed is what is corporeal, visceral, maternal and monstrous. These can include the sexual, oral and anal, all of which can be, and were, associated with the colonial. In Freud's 'The Uncanny' the crocodile emerges as a figure for the British colonial repressed to which he returns via the vehicle and vector of the artefacts of colonialism which bear the traces of other, alien or exotic places and peoples. The repressed does not have a fixed content or function but changes historically and varies culturally, though the crocodile and swamp as sites of the colonial unconscious have been with us for some time as we will see. Crocodiles for the ancient Egyptians were sacred, as they are for Australian Aboriginal peoples.

How the sacred becomes the monstrous, and even demonic, is a vexed topic. For David Quammen in a reading of the biblical Leviathan in *Job* 41 the monster is both created by God and subdued by Him in order to instil humility in His people (Quammen, 2004, p. 13). The monstrous and divine have been split off from each other and the divine constitutes itself and its hold over its believers by claiming to have created the monster, and, in turn, by subduing it. This splitting occurs in and as the moment of the institution of history. In pre-history the monstrous and divine are coterminous and mutually defining, and the uncanny is the name of this imbrication. Indeed, the monstrously sacred/demonic is perhaps an apt definition of the uncanny. Just as the sublime functions as a secular theology in which the sublime stands in for God in a culture for which God is dead (see Giblett, 1996, 2011), so the uncanny operates as a kind of secular demonology with the alligator and crocodile as Satan and the swamp as hell (as we have seen and will see below).

According to *The New Bible Dictionary* (Douglas, 1970, p. 729), Leviathan is usually considered to be 'some form of aquatic monster.' In *Psalm* 44: 26, 'It is clearly of the sea and is generally thought to be the whale,' though in the passage from *Job* 41, 'Most scholars agree that the creature is a crocodile.' The previous chapter of *Job* (40: 15–22) describes another monster translated as 'Behemoth' in the Authorised (or King James) Version. In the Revised Standard Version 'Behemoth' is footnoted as 'the hippopotamus.' The same version footnotes 'Leviathan' in the following chapter as 'the crocodile.' *The New Bible Dictionary* (Douglas, 1970, p. 138) concurs that 'Behemoth' in *Job* is 'the hippopotamus.' Yet the *New English Bible* (cited by Kelly, 2006, pp. 67 and 246) has 'crocodile' for 'Behemoth' in *Job* 40: 15 (this translation has 'whale' for 'Leviathan' in *Job* 41: 1). In this translation of the Bible the crocodile is both 'the chief of beasts' and 'the chief of God's beasts' (Kelly, 2006, p. 67). The crocodile is the king of the wet jungle, of the tropical wetland, the first, and best, work of God on the second day of creation (see Giblett, 2018, chapter 1).

Dragons, by contrast, may be the worst work of God, or at least the worst creation coming out of primeval chaos. In *Psalm* 74: 13–14 Leviathan is associated with 'the dragons of the waters.' Perhaps Parker (2015, p. 132) had this verse in mind when he described the most advanced Viking longships as leviathans. Their longships were not only dragon-prowed, but dragons themselves. Dragons can be aquatic creatures, but they can also be aerial, infernal and terrestrial creatures (see Giblett, 2018, chapter 3). In *Psalm* 74 Trudinger (2001, p. 35) comments that 'great importance is given to Earth [. . .]; as cosmos, it is a place of harmony and order. [. . .] The water-beings do not enjoy such a happy status. They are portrayed as God's opponents, suffer horrible injury and a degrading end.' Dragons and crocodiles are no exception. God is a dryland God, not a wetland God. He is a God of dryland agriculture and pastoralism. He is opposed to the wetland, its beings, and its aquaculture.

Leviathan has a much more ambiguous role and function in the Old Testament as a chaos monster and as a playmate than in the New Testament as the Devil or Satan (see Beal, 2002, p. 80). Whereas the dragon or Leviathan of the Old Testament is not associated with the Devil or Satan, in the last book of the New Testament the dragon or Leviathan is identified with the Devil or Satan (*Revelation* 9: 17 and 19; see Giblett, 2018, chapter 2). In the Old Testament, as Beal (2002, p. 81) comments, 'Neither Leviathan nor the sea monster nor any other monster is ever identified with the Devil or Satan.' In the last book of the New Testament, however, the dragon or Leviathan are identified with the Devil or Satan, and they are locked in the battle of 'good versus evil, God versus Devil' (Beal, 2002, p. 81).

Perhaps no animal has been more deified/demonized than the alligator and crocodile, the 'monarch of the marsh' and the 'king of beasts' of the tropical swamp. The alligatorian and the crocodilian have been repressed for a long time, at least since Freud's time, and it still persists. For Vollmar (1972, p. ix), 'Crocodiles, alligators and caimans both horrify and fascinate.' In Freud's (1985, p. 339) terms, they are uncanny as he defined the uncanny as 'what is frightening – what arouses dread and horror,' and I have defined the uncanny as what is both horrifying and fascinating (Giblett, 1996). Vollmar (1972, p. ix) suggests that 'lurid travellers' tales of evil reptiles lying loglike in tropical mud, ready to snatch and devour the unwary human, linger in the memory.' Freud (1985, p. 367, 2003, p. 151) developed the uncanny from reading one such tale, L. G. Moberly's 'Inexplicable,' published in *The Strand Magazine* in 1917. This story harks back to the ur-travellers' tale of alligators: William Bartram's account of his travels, and encounter with alligators, in a Florida swamp first published in 1791 (I return to both these stories shortly).

A century and a half later in the mid-nineteenth century John Muir walked a thousand miles from Kentucky to the Gulf of Mexico through a Florida swamp where he encountered an alligator. Fifty years later and a year before Moberly's story, Muir's account of his encounter with an alligator, like Bartram in a Florida swamp, was published posthumously. His reflections on this encounter are seen by some as a crucial transition in his thinking to a bio-centric ethic. More recently, Val Plumwood's (1996, 1999, 2000a, 2000b) gripping (perhaps the wrong word) autobiographical story of being crocodile prey published in (amongst other places) a collection by Travelers' Tales tries to avoid the lurid, but without success as I will argue later. For Plumwood, as with Muir, her experience was the occasion to reflect on her place in nature. If one is an uninvited guest in, or intruder into, the home of the alligator or crocodile, one can only expect to be prey as they are predators, or perhaps more precisely 'opportunist hunters and gorge feeders' (Kelly, 2006, p. 38). In other words, they are greedy feeders, not orally sadistic monsters.

In his reading of Moberly's story Freud downplays the role of real alligators and ignores the swamp as the place *par excellence* of the uncanny. If I were Freud,

I would psychoanalyze some long lost and repressed memories to do with his father and the phallus as the crocodile must be some sort of phallic symbol in Freud's lexicon of symbols, and to do with his mother and the swamp as a maternal place. Yet rather than psychoanalyzing Freud's psychopathology, I want to analyze the psychogeopathology that portrays the alligator and the crocodile as an orally sadistic monster, to engage in the talking cure of a psychoanalytic ecology that would regard them and the swamp in less demonic and more sacral terms, and to promote eco-mental health that would mean that these psychogeopathological symptoms did not arise in the first place (see Giblett, 1996, 2011).

Freud, perhaps in typical fashion, gives a three- or four-sentence summary of the story he 'came across' in a magazine. He does not give a reference other than saying that it appeared in 'a number of the English *Strand Magazine*' (Freud, 2003, p. 151). Some time ago I 'came across' two precise references to the story in Royle's *The Uncanny*, one to the original publication in *The Strand Magazine* and the other to a reprint in an anthology of stories from *The Strand Magazine* (Royle, 2003, pp. 140–141, n.3; Moberly, 1917, 1991). Rather than concentrating on the slips and mistakes in Freud's retelling as he himself would do, I want to consider the gaps and absences, the symptomatic lacunae, of Freud's reading in order to reinstate alligators and crocodiles living in a swamp as a vehicle and vector of the uncanny and disinvest their construction as orally sadistic monsters.

In 'The Uncanny' Freud (2003, p. 151) relates:

> During the isolation of the Great War, I came across a number of the English *Strand Magazine*. In it, among a number of pointless contributions, I read a story about a young couple who move into a furnished flat in which there is a curiously shaped table with crocodiles carved in the wood. Towards evening the flat is regularly pervaded by an unbearable and highly characteristic smell, and in the dark the tenants stumble over things and fancy they see something undefinable gliding over the stairs. In short, one is led to surmise that, owing to the presence of this table, the house is haunted by ghostly crocodiles, or that the wooden monsters come to life in the dark, or something of the sort. It was quite a naïve story, but its effect was extraordinarily uncanny.

Freud makes at least two mistakes in summarizing the story, the first of which is that the young couple move into a house, not a flat, furnished only with the table ('the table goes with the house' says the agent) and the table has carvings of alligators, not crocodiles, though Moberly's story itself slips between them as crocodiles, not alligators, inhabit New Guinea which is referenced in the story. These slips may be symptomatic of something more profound and if I were Freud I would no doubt think so and would analyze them (and him) for it. I am more interested, though, in the presences and absences in Freud's account, what he reproduces correctly and what he misses out altogether. Freud is not alone in referring to alligators or crocodiles as monsters as we have heard in a recent television report and as we shall see. There is a textual warrant for doing so in that the first-person narrator of the story refers to alligators as 'loathly monsters.' Yet this is not in relation to the carved, wooden alligators on the table, but in relation to a story within the story, a traveller's tale, about real, living alligators in a swamp in New Guinea that Freud does not consider at all as if he only read half the story by Moberly (more of that shortly).

First, the table. The first-person narrator (May, the wife of Freud's 'young couple') describes the effect the carved alligators had on her:

> As the light fell on the scaly bodies they had an extraordinary look of life, and the little sinister heads with the small evil eyes almost seemed to move. I shuddered and drew away from the table.
>
> (Moberly, 1917, p. 573, 1991, p. 184)

May feels quite faint for a moment and says to the agent, "There is such a queer smell in here,' [. . .] becoming all at once conscious of a strange and penetrating odour I had not before noticed.' She begins to suggest that it might be the drains but the agent is quick to interrupt and advise that they were "set in order before the last tenant vacated the house [. . .] I have the sanitary people's certificate about them" (Moberly, 1917, p. 573, 1991, p. 185). This terrace house is, however, not a working-class 'swamp' whose bad air was the supposed vector of malaria in the miasmatic theory of disease and the target of the Sanitary Movement (see Giblett, 1996 and its references). The carved figures on the table are creatures of the swamp and smell is a vector of the uncanny in the psychoanalytic theory of the psychopathology of everyday life in modernity (see Giblett, 1996).

May's husband Hugh's first encounter with the table is just as uncanny as hers. He runs his fingers over the carved surface of the table and rests them on the head of one of the alligators, 'a head fashioned with such skill that its loathsome naturalness made one shudder. 'Good heavens, May, the things look so lifelike I could almost have sworn one of them squirmed" (Moberly, 1917, p. 574, 1991, p. 186). Inanimate things coming to life, or seeming to do so, or imitating life, is for Freud one of the features and vectors of the uncanny.

The figuration of the alligator as 'a thing,' 'this thing' and 'the thing' persists into recent fiction as this is precisely how T. Coraghessan Boyle figures one in his 1990 novel *East Is East* (Boyle, 1990, pp. 335–336). 'The Thing' has been defined by Kristeva (1989, p. 13) as:

> the real that does not lend itself to signification, the centre of attraction and repulsion, the seat of sexuality from which the object of desire will become separated [. . .] the Thing is an imagined sun, bright and black at the same time.

'The Thing' occupies a zone prior to and outside signification, what Kristeva (1984) elsewhere calls 'significance,' which is also prior to and outside the separation of subject and subject in what Kristeva (1982) also calls abject. It is a creature of the uncanny, the centre and site of fascination and horror, the vehicle and vector of the monstrous.

Hugh and May are visited by a friend called Jack Wilding for whom the carved alligators trigger memories not related by Freud in his recounting of the story, which highlights the importance of the fact that the table is located in a house, not a flat. After dinner 'on a delicious May night' the two men friends are chatting with the windows open. The spring smells of the garden waft into the room:

> When all at once the drifting sweetness from without was tainted by that same strange odour which we had noticed once or twice before. . . . As it drifted across the room our guest suddenly sat bolt upright in his chair, and a curious greyness overspread his naturally bronzed complexion. 'My God!' he said, what is that? And why does it smell the same – the same –' His sentence trailed off into silence, and in the intense stillness following his strange words I heard a sound which, for some reason I could not pretend to explain, gave me a feeling of cold fear. I can only describe the sound as like a far-away bellowing – not precisely the bellowing of cattle, but a more sinister, more horrible sound, pregnant with evil. 'You hear it too?' Jack Wilding questioned, under his breath. . . . 'And the stench is here too! Good God! If I thought I should ever have to cross that swamp again I would go mad.'
>
> (Moberly, 1917, p. 576, 1991, p. 190)

The immediacy of the sense of smell takes Jack back to what was old and long familiar, and to what was long forgotten, if not repressed. Jack then recovers himself sufficiently to say:

> 'I must have had a nightmare – a waking nightmare,' he said, looking around him. 'I could have sworn that I smelt the alligator swamp in New Guinea, the place where –' he broke off short. 'I heard the loathsome brutes bellowing,' he began again; 'but, of course – or course, it was merely some association of ideas.'
>
> (Moberly, 1917, p. 577, 1991, p. 190)

Hugh indicates the table and suggests that it was the trigger for Jack's association of ideas:

> Jack turned and glanced at the table, and he recoiled when he saw the grinning heads lying amongst the crusted delicacy of leaves and flowers. 'Loathsome beasts!' he said, and again his voice shook. . . . 'I crossed an alligator swamp once with a friend. . . . It was dark, the place swarmed with those unspeakable devils, their stench was everywhere. It was dark – and poor old Danson' – he paused, as if speech were almost impossible – 'they dragged

> him off the path of the logs in the darkness.' . . . Somehow his words brought before me the hideous swamp, the darkness, the loathly monsters waiting for their prey, and the remembrance of just such an incident in a book I had once read flashed into my mind.
>
> (Moberly, 1917, p. 578, 1991, p. 191)

The original *Strand Magazine* publication of the story has accompanying illustrations by Dudley Tennant, one of which is of this story within the story, complete with a glimpse of both the table and the log path, and a depiction of Jack as both the teller of the tale and character in it (Moberly, 1917, p. 577).

Jack goes on to relate how 'the place swarmed with those unspeakable devils.' Swarming creatures, in biblical terms, are 'an abomination.' They are neither fish nor flesh nor fowl. They neither just swim nor walk nor fly, but do all three. These alligators are no exception. When the wooden alligators come to life they always seem to be sliding or slithering between Jack's or Hugh's feet (but not May's, as presumably she keeps her legs together like a lady). The housekeeper later describes how they go 'slithering' and 'running on their underneaths' (Moberly, 1917, p. 574, 1991, p. 193). They do not walk on all fours like domesticated animals. They do not separate their grotesque lower bodily stratum from the grotesque lower earthly stratum, but are part and parcel of it. The abominable is also almost beyond words. It is inexplicable as the story concludes and as its title signals.

Muir and the terror-stricken mouthful

A year earlier than the publication of Moberly's story, John Muir's *A Thousand Mile Walk to the Gulf* was published posthumously relating a journey he had undertaken almost 50 years earlier. In 1867 Muir, 'one of the seminal figures in the history of modern environmental thought' according to Frederick Turner in his introduction to the Penguin Nature Library edition of this book, encountered an alligator 'on the margin of a stagnant pool' in a Florida swamp which prompted Muir to reflect:

> These independent inhabitants of the sluggish waters of this low coast cannot be called the friends of man. [. . .] Many good people believe that alligators were created by the Devil, thus accounting for their all-consuming appetite and ugliness. But doubtless these creatures are happy and fill the place assigned them by the great Creator of us all. Fierce and cruel they appear to us, but beautiful in the eyes of God. They, also, are his children, for he hears their cries, cares for them tenderly, and provides their daily bread. The antipathies existing in the Lord's great animal family must be wisely planned, like balanced repulsion and attraction in the mineral kingdom. How narrow we selfish, conceited creatures are in our sympathies! How blind to the rights of all the rest of creation! With what dismal irreverence we speak of our fellow mortals! Though alligators, snakes, etc. naturally

> repel us, they are not mysterious evils. They dwell happily in these flowery wilds, are part of God's family, unfallen, undepraved, and cared for with the same species of tenderness and love as is bestowed on angels in heaven or saints on earth. I think that most of the antipathies which haunt and terrify us are morbid productions of ignorance and weakness. I have better thoughts of those alligators now that I have seen them at home. Honorable representatives of the great saurians of an older creation, may you long enjoy your lilies and rushes, and be blessed now and then with a mouthful of terror-stricken man by way of dainty!
>
> (Muir, 1992, pp. 98–99)

Here endeth the sermon on 'how I learned to stop hating alligators, and learnt to love them.' Only a true greenie would hug an alligator metaphorically as readily as they would a tree in actuality. Perhaps needless to say, Muir was not a mouthful of terror-stricken humanity for the alligators of a Florida swamp. He may have thought differently and not have been quite so jaunty if he had been as Val Plumwood was, and did, as we will see.

Bartram and the greedy monster

The story of the abominable crocodile and the monstrous alligator lurking in the uncanny swamp surfaces much earlier in modern Western culture than with Freud, Muir and Moberly in the early twentieth century. William Bartram in the late eighteenth century travelled to a Florida swamp where he encountered 'the subtle, greedy alligator' about to devour 'the voracious trout,' the eater eaten, the preyer predated by a larger predator:

> His enormous body swells. His plaited tail brandished high floats upon the lake. The waters like a cataract descend from his opening jaws. Clouds of smoke issue from his dilated nostrils. The earth trembles with his thunder.
>
> (Bartram, 1998, p. 75)

Two hundred years later Glasgow takes this last sentence as the subtitle for his social history of the American alligator and suggests that Bartram 'certainly wins, hands down, any contest for colourful alligator writing' (Glasgow, 1991, p. 31). There is certainly plenty of that in the intervening two centuries as Glasgow's study shows. Bartram's description is a lurid traveller's tale complete with exaggeration that the vapour is smoke for which he has a biblical warrant (*Job* 41: 20), though in a later encounter he describes how 'the vapour ascends from his nostrils like smoke' (Bartram, 1998, p. 82). It would only take the addition of fire from the alligator's mouth and wings on its back to complete the construction of the alligator as a dragon. Bartram's alligators are, as Slaughter (1996, p. 200) points out, 'monsters on a heroic scale. Bartram's drawing, *The Alegator* [sic] of *St Johns* [now housed in the Natural History Museum in London], resembles medieval representations of the dragons battled by St George.' Bartram on a couple of occasions refers to 'the monster' (Bartram, 1998, pp. 77 and 80). Bartram is a latter-day St George to the alligator's dragon (with some considerable biblical and cultural warrant for making this association (as we have seen previously)).

If, as I have argued elsewhere (see Giblett, 1996), the swamp is a secular underworld into which the hero of the modern adventure romance has to descend and in which he has to overcome monsters, then Bartram's travels involve him not only descending into Florida swamps but also overcoming their resident monsters, alligators – as he does later to emerge 'victorious, or at least [having] made a safe retreat' (Bartram, 1998, p. 80) as he concedes. Bartram leaves us in no doubt that we are in the underworld when he describes 'the boiling surface of the lake' (p. 75) created by two alligators engaged in 'horrid combat' (p. 76). The alligator swamp is a hellish, hot place and a place where the elements of water (lagoon) and fire (or heat) mix, as do the elements of air, earth and water created by the alligator itself. The 'dreadful roar' (p. 76) or the 'terrifying roar' of the alligator 'resembles very heavy distant thunder, not only shaking the air and waters, but causing the earth to tremble' (p. 82).

The alligator mixes the elements of earth, air, water and fire (thunder and lightning are the fire in the sky) just as the wetland does more generally (see Giblett, 2018, chapter 1; see also Giblett, 1996). Instead of these elements staying put in their proper place, the alligator and the wetland mix them up and violate the order of things that assigns them to a fixed and stable category. Even the smoke/vapour exhaling from the alligator's nostrils upsets the distinction between air, water and fire (where there is smoke, there's fire). The fact that the alligator in the water 'resembles, at a distance, a great chunk of wood floating about' (Bartram, 1998, p. 82) means that it pretends to be solid and earthy when in fact it is monstrous and slimy.

And orally sadistic to boot. If there is one feature of the alligator that more than any other arouses dread and horror for Bartram it is the fear of being 'dragged out of the boat and instantly devoured' (Bartram, 1998, p. 76) by the greedy monster. Its jaws are prodigious with the upper jaw opening at right angles to the lower one according to Bartram (p. 8 2) but mistakenly for Glasgow (1991, p. 31). Bartram also found that they are capable of 'belching floods of water' (Bartram, 1998, p. 76). The jaws of the alligator emit terrifying roars and belch water, and so give out noise and liquid. They also take in liquid in the form of water and blood, and solid in the form of flesh, animal or human. Their jaws give and take. Their jaws are a transportational zone between the inside and outside of the body through which matter passes.

Alligators in a Florida swamp function in both Bartram's and Muir's texts as figures for the British colonial and American cultural unconscious. Florida and the American south more generally with its swamps have been repressed in the collective psyche of the American north and have functioned as a figure of the primitive and backward (see Giblett, 1996 and references). Relatedly, American wetlands have been drained and their creatures commodified from early colonial times. In Bartram's case he is returned to this repressed only to try to continue to repress it; in Muir's, he embraces it – albeit metaphorically.

Plumwood and great-toothed jaws

Unlike both Muir and Bartram and their encounters with an alligator in a Florida swamp, but like Danson in Moberly's story, the Australian ecofeminist and eco-philosopher Val Plumwood was a mouthful of terror-stricken humanity for a crocodile in Kakadu National Park in northern Australia in 1985. She lived to tell the tale, though, unlike Danson, but it took some time for her to be able to tell it. I am not going to attempt to retell the story in all its details and if I attempted to sum it up by saying it was a gripping and gut-wrenching story I would be making bad puns. Plumwood's story is worth reading for itself. By reading it critically I am not belittling or demeaning her experience, which was traumatic to state the obvious, nor her individually as she was heroic to say the least. Rather, as with Freud, I am trying to critique the patriarchal and colonial elements that emerge in it despite her best efforts to keep them at bay.

Her story has some explicit elements of the uncanny, such as her description of 'the unfamiliar sensation of being watched' and her 'whispering sense of unease' prompted, not by the sight of a crocodile, but of 'a strange rock formation' (Plumwood, 1996, pp. 33–34, 1999, p. 78, 2000a, p. 57, 2000b, p. 130) which is a portent of what is to come. Plumwood hears whispers of unease and observes the strange rock formation so she is in a heightened state of sensory alertness. When the crocodile attacks her flimsy fibre-glass canoe 'the unheard of was happening' (Plumwood, 1996, p. 34, 1999, p. 78, 2000a, p. 131, 2000b, p. 57). The uncanny is evoked by hearing (or not) and smelling, by what can be heard or not, what can be smelt, but generally not by seeing. Deathly silence evokes the uncanny. In this case, being seen also evokes it. 'Being prey' as she calls it, is the result of being seen. To be prey one has been seen already. Plumwood is not the subject of the gaze, but its object. She hears whispers of unease and hears (and sees) the unheard of, a crocodile attacking a canoe.

The ten-year gap between 'being prey' and publishing her account and reflections on this event under this title attests perhaps to the difficulty for her of coming to terms with her experience and expressing it. Part of the difficulty was not only the trauma of the attack itself but also the way in which her story was subjected to what she called 'the cultural drive to represent it [the attack] in terms of the masculinist monster myth: the master narrative' (Plumwood, 1996, p. 40, 1999, p. 85, 2000a, p. 139, 2000b, p. 59). In this myth, the crocodile is constructed as a ravening, orally sadistic monster who (or which) rapes and eats his (and it is always a male in the myth) innocent female victim. Yet Plumwood reproduces the myth herself when, for example, she earlier describes in lurid terms how she had 'a blurred, incredulous vision of great toothed jaws bursting from the water' that then 'seized [her] between the legs in a red-hot pincer grip' (Plumwood, 1996, p. 35, 1999, p. 79, 2000a, p. 131, 2000b, p. 57). The crocodile is figured as a monstrous, orally sadistic and reptilian cousin of 'Jaws,' the monstrous shark.

Rather than only construing her experience in good ecological terms of being prey and so being a part of the food chain, Plumwood also turns in the longer version of her story to the mythology of New Guinea (and why not that of Australian Aborigines, one wonders?) as a way of accounting for her experience that does not merely deny or repress its mythological elements and significance. She suggests that 'crocodiles are masters of water' and goes on to argue that:

> the crocodile is an exploiter of the great planetary dualism of land and water. As Papua New Guinea writer Vincent Eri suggests in his novel, *The Crocodile*, the creature is a sort of magician: its technique is to steal the Other, the creature of the land, away into its world of water where it has complete mastery over it. Water is the key to the crocodile's power, and even large crocodiles rarely attack in its absence. The crocodile is then a boundary inhabitant.
>
> (Plumwood, 1996, p. 39, 1999, p. 84, 2000a, p. 137)

In other words, the crocodile, like the alligator, is a wetland inhabitant, an inhabitant of the intermediary zone between dryland and deep water that crosses 'the boundary' between land and water and upsets the dualism between them (see Giblett, 1996). Yet, like Freud who ignores the story within the story of the alligator swamp in New Guinea (and like Royle who ignores the alligators and the swamp in the story within the story), and like both who ignore these aspects as the vector and vehicle for the uncanny, Plumwood overlooks the fact that the wetland with its distinctive features is the crocodile's habitat. She describes the latter as 'the swamp' and contrasts that with 'Kakadu's wetlands' with their 'dreamlike beauty' enticing her into 'a joyous afternoon's idyll' at the beginning of the story. She describes how here she 'glutted' herself on 'the magical beauty [. . .] of the lily lagoons untroubled by crocodiles' (Plumwood, 1996, p. 33, 1999, p. 76, 2000a, p. 128, 2000b, p. 56). She moralizes the pastoral world of the wetland as good and heavenly – good enough to eat with herself as metaphorical glutton – and the swamp as bad and hellish, the place of the orally sadistic and gluttonous crocodile. For her, the crocodile lives in a swamp whereas the tourist visits a wetland; in brief, swamp is bad, wetland is good.

Naturally, she would rather eat than be eaten (and wouldn't we all?). Just as she is being watched, but prefers to watch, so she prefers to 'eat' the beauty gluttonously than to be eaten greedily by the crocodile. Who can blame her? The point is, though, oral sadism rules, okay? She is a metaphorical glutton and the crocodile is figured as an orally sadistic monster. Tourist and crocodile are going about their normal, everyday business subscribing in both cases to the master narrative – despite Plumwood's best efforts to avoid it and do otherwise. The tourist watches (preferably without being watched) and consumes beauty gluttonously through his or her eyes without being consumed; the crocodile watches the tourist and sometimes consumes him or her through his or her mouth. Both are just doing their own thing in this mythology of modernity with its non-reciprocal and non-symbiotic positions of mastery of watching and eating, rather than being watched and eaten.

As the Kakadu wetland for Plumwood is good and heavenly, the crocodile swamp by implication is bad and hellish, though the wetland and the swamp are one and the same place – they are just figured differently. The Kakadu wetland is Eden before the Fall, or before the appearance of the serpent, here transformed into the crocodile. It is also the place of good magic, but crocodiles are excluded from this world of good magic as they bring trouble into paradise by enticing creatures of the land with bad magic into a watery grave, into a wet underworld, into its world – the swamp. The good magic is a sanitized white magic enticing Plumwood into the beautiful pastoral idyll of the wetland split off from, and valorized over, the black magic of the crocodile enticing her into the horrifying black water of the swamp. Magic is just magic, though, without the moralization. Plumwood (curiously for an ecofeminist) reproduces not only the patriarchal, Western moralization of the wetlandscape, but also its dualisms and spatial metaphysics and poetics of land and water, good and bad, white and black, heaven and hell, above and below (see Giblett, 1996).

Yet humans are not simply or exclusively creatures of the land, just as crocodiles are not simply or exclusively creatures of the water. Alligators and crocodiles are creatures who live in the two elements of earth and water, just as their wetland home mixes these elements (Strawn, 1997, p. 14, Giblett, 1996). Humans are also creatures of water as we are predominantly made up of water. Our beginnings as individuals are in the watery world of the womb and our beginnings as a species in evolutionary terms are in the womby world of water (see Giblett, 2018, chapter 1; see also Giblett, 1996). Humans also have vestigial reptilian parts of the brain. Humans are meat for crocodiles (and vice versa), but we are also very distant cousins. We are both wetland creatures living on a complementary, non-dualistic planet of land and water, in short, of waterland.

Alligators and crocodiles in Haraway's (2008) terms are 'companion species' (though not in the same way as pets are) who occupy what she also calls a 'contact zone' between human beings and non-human beings (though again not in the same way as the human home, or domestic space, is for pets). The quaking zone is a contact zone between humans and non-humans (such as alligators and crocodiles in native quaking zones, and lice and rats in feral quaking zones of the trenches and of the ruined cities of world warfare and of urban slums). This contact zone can be phatic involving communication via the acts and senses of touching, smelling, hearing and seeing, or it can be phagic involving consumption via the acts of biting, chewing, etc. and the sense of taste. The phatic contact zone as a rule occurs with pets; the phagic with non-pets. If the pet becomes phagic, it ceases to be pet (and phatic) and may be 'put down.'

The uncanniness of alligators and crocodiles lies in the fact that they and modern urban humans live in the same earth-home (ecosphere), but not in the same house. Modern urban humans may live in the earthly wetland home of alligators and crocodiles, but the latter do not live in human houses and gardens as a rule. Alligators and crocodiles are (un)homely creatures, like human beings, of the terraqueous globe as Serres (1997, p. xiii) calls the earth. For Haraway (2008, p. 45), 'The familiar is always where the uncanny lurks. Further, the uncanny is where value becomes flesh again.' The familial – the family of 'Man,' the family of reptilians (including humans) – is where the alligatorian and crocodilian lurk just beneath the surface of the wetlands of the earth. Flesh is what carnivores – human, alligatorian and crocodilian – eat. The body becomes flesh in the uncanny; the value of the body is chomped into the flesh of a 'mouthful of terror-stricken humanity' as for Muir and Plumwood in the jaws of an alligator or crocodile, or of a mouthful of crocodile burger in my jaws as when I visited a crocodile farm outside Lusaka in Zambia in 2007. Eat or be eaten indeed.

Plumwood draws on Eri's novel to support her argument about crocodiles and water, but her reading has no real basis or textual warrant. Beside the perhaps customary references in the novel to 'the monstrous crocodile' and to 'the horrifying creature' (Eri, 1973, pp. 108 and 113), the crocodile of the title does not figure (in two senses of the word) much in the novel. It is certainly not used, as Plumwood later suggests (1996, p. 39, 1999, p. 85, 2000a, p. 138), as:

> a metaphor for the relationship between colonised indigenous culture and colonising Western culture. If the crocodile-magician-coloniser can drag you completely into its medium, you have little chance; if you can somehow manage to retain a hold on your medium, you may survive.

If the crocodile is used in Eri's novel as a metaphor for anything, it is as a device for explaining the inexplicable, unlike for Moberly for whom the crocodile *is* the inexplicable. When Mitoro, the wife of the central character Hoiri, disappears the whole village maintains that a crocodile took her. Interestingly in one Aboriginal story the crocodile is a wife-stealer too (see Mudrooroo, 1994, p. 33). The men of the village hunt down a crocodile and Hoiri kills it in an act of revenge that he is privileged to perform as the victim. Yet at the end of the novel Hoiri encounters Mitori. She does not acknowledge him and no explanation for her leaving him is given other than that she is under the power of the magicians who also control or transform themselves into crocodiles.

The crocodile is a scapegoat onto which her sin of leaving her husband is heaped (as she cannot be found and punished) and the village is expiated. The crocodile is a creature of the wetland figured as the scapeland, not only in the sense of the anti- or counter-landscape but also in the sense of the sacrificial victim onto, or into, which the sins of the community are heaped and expiated (see Giblett, 1996). Perhaps in modern Western medical terms Mitoro was suffering from post-natal depression following the birth of their son and left Hoiri as a result. This would highlight the connection between melancholia and wetlands, and the creatures of the wetland (see Giblett, 1996). Alternatively, perhaps she simply ran away with another man. The crocodile-magician is a part of indigenous culture that is not necessarily good, or acts for good, but functions to explain the inexplicable unlike in Moberly's story of this title.

The colonizer, on the other hand, is largely represented by piggish Patrol Officers or draconian District Officers. One of them is 'referred to as "the crocodile," a title that was one of praise rather than abuse' (Eri, 1973, p. 141), presumably because of his power, cruelty and ugliness. He is monstrous and horrifying like the crocodile. The colonizer certainly wants to drag the indigine into his medium represented by patrols, prisons, cities and warships. The crocodile does not represent this world. It represents an unsettling liminal zone of indigenous culture that is not necessarily good or bad, but it is certainly not the colonizer's culture, or a metaphor for it other than for the monstrous, horrifying and inexplicable in it, and all cultures, as in Moberly's story.

Plumwood associates the crocodile with the relationship between colonizer and colonized; Freud associates the uncanny with an artefact of colonialism with carved crocodiles that seem to, or do, come alive. Both are associated with a New Guinea swamp. The crocodile surfaces uncannily here in an Australian ecofeminist text quoting a novel, a colonial genre, written in a colonized culture. It also emerges in Freud's Viennese cultural and psychoanalytic repressed via a British magazine as a vector for the uncanny. In both texts the crocodile surfaces as a figure for the British colonial unconscious – repressed, but returning in their slips, gaps and lacunae via the vehicle and vector of the artefacts of colonialism which bear the traces of other, alien or exotic places and peoples.

In Beth Yahp's 1992 novel, *The Crocodile Fury*, set in Malaysia the crocodile functions in a slightly different way as a projection (or 'symbol') of repressed desires and fears. Male sexual aggression for the first-person narrator's Grandmother is represented by 'the land crocodile,' 'a creature who can't be controlled.' For her:

> This beast, this human-shaped terroriser of innocent girls [. . .] lives on the edges of the jungle so he gets the best of both worlds [. . .] when the urge takes him, out he creeps to jump at victims. [. . .] When he's on a rampage his whole body swells, he lets out a sound that is low and throbbing, that fills young girls' and old women's heads with the hum of a thousand insects, the shudder of earthquakes, the toppling of great jungle trees, so they fall over faint and panting. When the crocodile is on a rampage he gives off a red glow. When a young girl is deflowered or a man seduces some else's wife, or your daughter is stolen, your faithful assistant led astray, you can be sure it's with the help of the land crocodile. A girl unlucky enough to be touched by the crocodile will forever be branded with wild ideas, with tempestuous fevers. She'll never listen to her elders, she'll never do what she's told. Such a girl will become a lifetime victim of passions so itchy she'll never be able to sit still.
>
> (Yahp, 1992, p. 18)

Redolent of Bartram's alligator, the land crocodile is a creature of the quaking zone of earthquakes and of the liminal zone, not between land and water, but between city and jungle. He crosses the boundary between the two to get the best of both worlds. Like Eri's crocodile, the land crocodile is a convenient explanatory device, a heuristic trope, for explaining inexplicable, abnormal, or unusual, or typical male sexual behaviour.

Monstrosity to mutuality

My three encounters or interactions with alligators and crocodiles have been neither as spectacular nor as close as Plumwood's, and pretty tame by comparison, but instructive nevertheless along the lines I have been pursuing in this chapter. I visited the Kimberley in 1992 and ventured into salt-water crocodile country down the lower Ord River in a small boat operated by a tour guide. This venture was part of research for a book on Western Australian wetlands with Hugh Webb. The photographer accompanying us, Simon Neville, took photographs of crocodiles which involved getting within a few metres of the shore. Greg, the guide, pointed out that he only went this close when there was enough water under the boat so that if the croc decided it wanted to go into the water it could go under the boat, rather than launching itself across the top of the boat and taking one of us with him or her. We all appreciated this consideration of *not* being prey!

My other interactions with alligators and crocodiles were perhaps a bit tamer and much more mainstream touristic. My most recent interaction with crocodiles was in 2016 in Costa Rica on a guided tour of the famous Tarcoles River. The boat trip included a sighting of the biggest crocodile on the river called 'Mike Tyson,' which probably had more to do with his size than his namesake's conviction for rape. I couldn't help wondering about Plumwood's story though. The captain of the boat managed to herd Mike towards the shore at which point the guide jumped out of the boat into the water and coaxed Mike towards the shallows by waving a yellow cloth. Mike came closer while the guide continued to wave the cloth. Mike obligingly opened his mouth and the guide got the camera of one of my fellow tourists and took a photo a few inches from Mike's nose. I asked the captain of the boat if the guide got paid danger money. Mike must have been well fed, or unthreatened, or familiar with the guide, though he said that they hadn't seen Mike for a while and he had obligingly shown up for our visit.

My one and only interaction with alligators was perhaps a bit tamer and much more mainstream touristic. Like Bartram and Muir, it was in a Florida swamp. Wakulla Springs is a popular picnic and swimming spot outside of Tallahassee. According to the inevitable brochure for the place, Wakulla is a Seminole Creek word that means 'strange and mysterious waters,' in a word, uncanny, but not monstrous, like the alligators. Such films as *Creature from the Black Lagoon* and some of the old Tarzan films with Johnny Weissmuller were shot there. You get the general picture of swampy jungle. It is alligator habitat, too, so the alligators are supposed to stay on one side of the river and people on the other. A lifeguard is stationed on the people side and if an alligator ventures across to this side the lifeguard evidently yells out, 'Alligator on the people side!' Everyone is supposed to get out of the water. The people side is roped off to stop people going across to the alligator side. A short boat tour takes tourists downstream and into alligator territory for a closer look. I saw two alligators and two turtles lying on the same log. Perhaps this is an instance of mutuality with two species living together and sharing the same living space, just as alligators and humans share the swimming hole with a degree of respect on the part of humans for the alligators in this place, but not in many others (see Strawn, 1997, esp. p. 170; Giblett, 2011). Alligators, crocodiles and humans should be able to live together in a psycho-symbiotic relationship and pursue their livelihoods in the same bioregion on the same earth.

References

Bartram, W. 1998. *Travels: Naturalist's Edition.* F. Harper, ed. Athens, GA: University of Georgia Press. First published 1791.

Beal, T. 2002. *Religion and Its Monsters.* New York: Routledge.

Boyle, T. 1990. *East Is East.* New York: Penguin.

Douglas, J., ed. 1970. *New Bible Dictionary.* London: Inter-Varsity Press.

Eri, V. 1973. *The Crocodile.* Ringwood, Victoria: Penguin.

Freud, S. 1985. The 'Uncanny.' In: *Art and Literature.* J. Strachey, trans., Penguin Freud Library 14. Harmondsworth: Penguin, pp. 335–376. First published 1919.

Freud, S. 2003. *The Uncanny.* D. Mclintock, trans. London: Penguin. First published 1919.

Giblett, R. 1996. *Postmodern Wetlands: Culture, History, Ecology.* Edinburgh: Edinburgh University Press.

Giblett, R. 2011. *People and Places of Nature and Culture.* Bristol: Intellect Books.

Giblett, R. 2016. *Cities and Wetlands: The Return of the Repressed in Nature and Culture.* London: Bloomsbury.

Giblett, R. 2018. *Environmental Humanities and Theologies: Ecoculture, Literature and the Bible.* London: Routledge

Glasgow, V. 1991. *A Social History of the American Alligator: The Earth Trembles With His Thunder.* New York: St Martin's Press.

'Gran Who Beat Off Croc Attack' 2004. *The West Australian*, 16 October, p. 40.

Haraway, D. 2008. *When Species Meet.* Minneapolis: University of Minnesota Press.

Kelly, L. 2006. *Crocodile: Evolution's Greatest Survivor.* Crows Nest, NSW: Allen and Unwin.

Kristeva, J. 1982. *Powers of Horror: An Essay on Abjection.* L. Roudiez, trans. New York: Columbia University Press.

Kristeva, J. 1984. *Revolution in Poetic Language.* M. Waller, trans. New York: Columbia University Press.

Kristeva, J. 1989. *Black Sun: Depression and Melancholia.* L. Roudiez, trans. New York: Columbia University Press.

Laplanche, J. and Pontalis, J.-B. 1973. *The Language of Psychoanalysis.* Donald Nicholson-Smith, trans. London: Hogarth.

Makari, G. 2008. *Revolution in Mind: The Creation of Psychoanalysis.* New York: Harper.

Moberly, L. G. 1917. Inexplicable. *The Strand Magazine*, 54(324), pp. 572–581.

Moberly, L. G. 1991. Inexplicable. In: *Strange Tales From the Strand.* J. Adrian, ed. Oxford: Oxford University Press, pp. 183–195.

Mudrooroo. 1994. Crocodiles. In: *Aboriginal Mythology: An A-Z Spanning the History of Aboriginal Mythology From the Earliest Legends to the Present Day.* London: HarperCollins, pp. 33–35.

Muir, J. 1992. *A Thousand-Mile Walk to the Gulf.* New York: Penguin. First published 1916.

Parker, P. 2015. *The Northmen's Fury: A History of the Viking World.* London: Vintage.

Plumwood, V. 1996. Being Prey. *Terra Nova*, 1(3), pp. 32–44.

Plumwood, V. 1999. Being Prey. In: *The New Earth Reader: The Best of Terra Nova*. D. Rothenberg and M. Ulvaus, eds. Cambridge, MA: MIT Press, pp. 76–91.

Plumwood, V. 2000a. Being Prey. In: *The Ultimate Journey: Inspiring Stories of Living and Dying*. J. O'Reilly, S. O'Reilly and R. Sterling, eds. San Francisco: Travelers' Tales, pp. 128–146.

Plumwood, V. 2000b. Being Prey. *UTNE Reader*, July–August, pp. 56–61.

Quammen, D. 2004. *Monster of God: The Man-Eating Predator in the Jungles of History and the Mind*. London: Hutchinson.

Roudinesco, É. 2014. *Sigmund Freud en son Temps et dans le Nôtre*. Paris: Seuil.

Roudinesco, É. 2016. *Freud: In His Time and Ours*. C. Porter, trans. Cambridge, MA: Harvard University Press.

Royle, N. 2003. *The Uncanny*. New York: Routledge.

Serres, M. 1997. *The Troubadour of Knowledge*. S. Glaser with W. Paulson, trans. Ann Arbor: University of Michigan Press.

Slaughter, T. 1996. *The Natures of John and William Bartram*. New York: Random House.

Strawn, M. 1997. *Alligators, Prehistoric Presence in the American Landscape*. Baltimore: The Johns Hopkins University Press.

Trudinger, P. 2001. Friend or Foe? Earth, Sea and *Chaoskampf* in the Psalms. In: *The Earth Story in the Psalms and the Prophets*. N. Habel, ed. Sheffield: Sheffield Academic Press, pp. 29–41.

Vollmar, F. 1972. Preface. In: *Crocodiles: Their Natural History, Folklore and Conservation*. C. Guggisberg, ed. Mount Eliza, Victoria: Wren, pp. ix–x.

Whatmore, S. 2002. *Hybrid Geographies: Natures Cultures Spaces*. London: Sage.

White, T. 1997. Polk Salad Annie. *Tony Joe White Collection* [CD]. Festival Records.

Whitebook, J. 2017. *Freud: An Intellectual Biography*. Cambridge: Cambridge University Press.

Yahp, B. 1992. *The Crocodile Fury*. Sydney: Angus and Robertson.

3 The uncanny urban underside

For some late nineteenth-century writers, such as Engels, Gissing, London and Thomson, the modern city has a dark underside figured as stagnant pool, bottomless abyss, nether world or dreadful night. The commonplace responses of dread and horror that were projected on to the native quaking zone were displaced on to the industrial quaking zone of the urban underside. This chapter argues that these writers are 'placist' and exercise placial discrimination of one privileged place against another denigrated (literally 'blackened') place. In this case, they ascribe characteristics to a (man-made) place (the city) that were previously ascribed to a place not made with human hands (the jungle, abyss, nether-land of swamps, etc.). The negative connotations that attach initially to the native quaking zone are attached subsequently to the feral quaking zone.

By contrast, for some early twentieth-century writers, such as Freud and Walter Benjamin, the underside of the urban is both fascinating and horrifying, and figured in bodily terms. Recently for some early twenty-first century writers the back blocks and slums of the hypermodern city – characterized as regions of rust and ruin, or of decay and decomposition steeped in feral swamps – are fascinating. These twenty-first century writers express anxiety or dread about the city like their nineteenth-century counterparts. Rather than this fraught figuration, I propose remythifying the city as a body, not only with parks as lungs as in the nineteenth-century cliché of the city, but also with twentieth-century skyscrapers as the head and brain; the water supply and sewerage systems as the oesophagus and intestines; the rivers as arteries; the wetlands as kidneys, liver and placenta; the 'mouth' and the estuary of the river as the anus and bowels; slums as a sore to be treated and cared for, not bled or lanced in the gentrification of slum clearance; and so on. Rather than figuring slums as swamps in pejorative terms as places of disease and horror, I propose seeing them as artificial wetlands whose livability for their residents would be improved by improving their ecological functionality as kidneys, liver and placenta, as places of hope and new life.

It is a commonplace of detective fiction that the modern city has a dark underside. Arthur Conan Doyle refers to 'the dark jungle of criminal London' in one of his Sherlock Holmes stories. The role of the private detective is to slash through this tropical growth, penetrate the darkness and bring enlightenment to its benighted citizens. For other late nineteenth-century writers the dark underside of the city looms large in slightly different, though similar, ways. For Jack London it is the abyss; for Rudyard Kipling and James Thomson it is 'The City of Dreadful Night'; for George Gissing it is the nether world; and for Friederich Engels it is a stagnant pool. This figuration and association persists into some of the most recent writing about the city, such as Antoine Picon's reading of the anxious landscape of the rusting city. How this dark underside is figured adumbrates the politics, pleasures and fears of city places and urban spaces.

These cognate views of the dark underside of the urban as nether land contrast with two early twentieth-century writers, Walter Benjamin and Sigmund Freud, especially when it comes to the lower-class inhabitants of the urban underside, such as servant girls and prostitutes. For Gissing the nether world of servant girls and prostitutes was exclusively an object of horror, whereas for Benjamin it was a source of fascination. For Freud it evoked both horror and fascination as it was a place where the uncanny was manifested. Yet despite their differences, for all these writers on the urban its dark underside is experienced as a psychological region of dread, anxiety or the uncanny.

The darkness of the modern city is constituted not only by its immoral underside, its dens of thieves, its parade of prostitutes, its labyrinthine slums, but also by the physical and moralized underside of the open drains or closed sewers that service it and the swamps on which it may, more than likely, have been built. The latter is often employed to figure the former with the *demi-monde* of the slums figured as swamp, or sewer, or muddy shore. Mayhew's aim with what he calls 'the neglected class' is to try to 'lift them out of the moral mire in which they are wallowing' (Mayhew, 1985, p. 41). Mayhew's source for his miry metaphor is his own description later of the 'mud-larks' who 'plash their way through the mire' (p. 210) and wade 'through the mud left on the shore by the retiring tide' and whose 'bodies are grimed with the foul soil of the river' (p. 209). Both sides of the figuration come together in the uncanny and the nether world (or nether lands).

Uncanny urban underside

The uncanny for Freud is a feeling or state of fascination and horror evoked by the 'dark continent,' whether it be of female sexuality, of the slums or the swamp, or its monstrous creatures, such as alligators or crocodiles (as we saw in the previous chapter). Freud (2003, p. 144) relates in 'The Uncanny' how he was:

> strolling one hot summer afternoon through the empty and to me unfamiliar streets of a small Italian town. I found myself in a district about whose character I could not long remain in doubt. Only heavily made-up women were to be seen at the window of the little houses, and I hastily left the narrow street at the next turning. However, after wandering about for some time without asking the way, I suddenly found myself back in the same

> street, where my presence began to attract attention. Once more I hurried away, only to return there again by a different route. I was now seized by a feeling that I can only describe as uncanny, and I was glad to find my way back to the piazza that I had recently left and refrain from any further voyages of discovery.

Freud's lost object is himself that he repeatedly finds in the wrong place, or more precisely, the place to which he unintentionally returns. This return enacts an unconscious desire. The uncanny is a return *to* the repressed.

In Freud's autobiographical anecdote the repressed is not only his sexual repressed, but also the morally and spatially repressed of the small Italian town, its red-light district. The sexually and spatially repressed is the bodies of the prostitutes embodying what Solnit (2000, p. 209) calls 'transformation of city into female body.' For the young Walter Benjamin (1979, pp. 330–331, 1999, p. 623) the city of Berlin is transformed into the body of servant girls. He relates:

> The dream ship that came to fetch us on those evenings must have rocked at our bedside on the waves of conversation, or under the spray of clattering plates, and in the early morning it set us down on the ebb of the carpet beating that came in at the window with the moist air on rainy days and engraved itself more indelibly in the child's memory than the voice of the beloved in that of the man – this carpet beating that was the language of the nether world of servant girls, the real grownups, a language that sometimes took its time, languid and muted under the grey sky, breaking at others into an inexplicable gallop, as if the servants were pursued by phantoms. The courtyard was one of the places where the city opened itself to the child.

The languid and muted language of the servant girls is not only their spoken language but also the body language of their carpet-beating that was vaguely arousing and sexually enticing for the young Walter with its overtones of sadism on the part of the servant girls and masochism on Benjamin's. The courtyard, rather than being an entry into the private domestic space of the home, was a passage going out into the public space of the city, and into the female body culminating in his later depiction of, as Solnit (2000, p. 209) puts it, 'Paris as labyrinth . . . whose centre is a brothel.' The city for Benjamin is ultimately a labyrinthine female body (as discussed in chapter 5).

Nether world

Benjamin's fascinating nether world of servant girls is worlds away from the horrifying sub-terrain of George Gissing's 1889 novel, *The Nether World*. Benjamin's courtyard where the city opened itself to the child is a different place to Gissing's courtyard where the city closed in upon itself to imprison and crush the individual. The moist air on rainy days of a lovingly evoked childhood in Benjamin's case is an ocean apart from the description in Gissing's novel of 'a rank, evilly-fostered growth' like a poisonous swamp plant in contemporaneous representations of swamps (see Giblett, 1996). His narrator is not referring to a plant here though. He is comparing one of the working-class girl characters, 'not to some piece of exuberant normal vegetation,' but to a kind of festering vegetative sore on the backside of the body-politic. Yet their labour is necessary for the functioning of the body-politic which could not survive without it.

The narrator offers the opinion that 'the putrid soil of that nether world yields other forms besides the obviously blighted and sapless' (Gissing, 1992, p. 8). Indeed, it yields the bad smelling and the morally bad. This is hardly surprising when 'filth, rottenness, evil odours, possessed these superfluous dens of mankind and made them gruesome to the peering imagination' (p. 74). Into 'the jaws of this black horror' the narrator, or at least his imagination, has peered. The nether world is figured as monster and the narrator as dentist who peers (albeit in imagination, not in reality) into its orally sadistic jaws. The nether world is a living creature to be shunned as an abomination. The narrator leaves us in no doubt about this when he describes 'a disagreeable quarter, a street of squalid houses, swarming with yet more squalid children' (p. 129). These swarms of children have developed from a 'swarm of babies' (p. 132). According to the Levitical interdiction, swarming creatures are an abomination and are to be abhorred (see Douglas, 1966; Giblett, 2018).

The swarming nether world of the city with its vertiginous depths is worlds away from the pastoral upper world of the country with its flat horizons (though it too has its dark side). This contrast comes into stark relief when making a journey by railway from the city to the country:

> Over the pest-stricken regions of East London, sweltering in sunshine which served only to reveal the intimacies of abomination; across miles of a city of the damned, such as thought never conceived before this age of ours; above streets swarming with a nameless populace, cruelly exposed by the unwonted light of heaven; stopping at stations which it crushes the heart to think should be the destination of any mortal; the train made its way at length beyond the outmost limits of dread, and entered upon a land of level meadows, of hedges and trees, of crops and cattle.
>
> (Gissing, 1992, p. 164)

The unhomely and unhealthy artificial wetland of the city for Gissing gives way to the homely and healthy natural dryland of the country:

> It is merely one of those quiet corners of flat, homely England, where man and beast seem on good terms with each other, where all green things grow in abundance, where from of old tilth and pasture-land are humbly observant of seasons and alternations, where the brown roads are familiar only with the tread of the labourer, with the light wheel of the farmer's gig, of the rumbling of the solid wain.
>
> (Gissing, 1992, pp. 164–165)

This is an idealized picture straight out of Constable of an organic community with its pre-industrial technologies living in pastoral harmony in accord with seasonal rhythms and where the growths are green and abundant, not rank and evilly festered, nor blighted and sapless. It nevertheless has its dark side as Gissing goes on to relate:

> Here, as elsewhere, the evil of the times was pressing upon men and disheartening them from labour. Farms lying barren, ill-will between proprietor and tenant, between tenant and hind, departure of the tillers of the soil to rot in towns that have no need of them.
>
> (p. 165)

The normal, fertile country has been blighted and the symptoms of rural depopulation are blamed on the age rather than its causes in industrial capitalism diagnosed. The country still has its consolations, though, as a salve for the deprivations of the city:

> Danbury Hill, rising thick-wooded to the village church, which is visible for miles around, with stretches of heath about its lower slopes, with its far prospects over the sunny country, was the pleasant end of a pleasant drive.
>
> (p. 165)

The church is a symbolic beacon on the hill in dark times, a symbol of the city of God on high, and a light at the end of the tunnel to which the traveller journeys and from which the viewer can command a pleasing prospect of the country. This pleasing prospect is unlike the displeasing prospects of the city where the traveller walks 'through all the barren ways and phantom-haunted refuges of the nether world' (p. 247). Both country and city have become barren, but at least the country is not the city of the dammed and the dead. 'Mad Jack' declares, "This place to which you are confined is Hell. . . . This is Hell – Hell – Hell!" (p. 345). Similarly in Morrison's (1996, p. 2) *A Child of the Jago*, published in 1896, the rejoinder to the stock epithet 'Go to hell,' uttered by one character, is, "Hell? You're in it! . . . There can be no hell after this," uttered by another.

The pleasing prospect from Danbury Hill is worlds away from Shooter's Gardens whose 'walls stood in perpetual black sweat; a mouldy reek came from the open doorways; the beings that passed in and out seemed soaked with grimy moisture, puffed into distortions, hung about with rotting garments' (Gissing, 1992, p. 248). The verdant horizontal surface of 'England's green and pleasant land' is contrasted with the vile, vertical walls of the Farrington Road buildings:

> Vast, sheer walls, unbroken by even an attempt at ornament; row above row of windows in the mud-coloured surface, upwards, upwards, lifeless eyes, murky openings that tell of bareness, disorder, comfortlessness within. . . . An inner courtyard, asphalted, swept clean – looking up at the sky as from a prison. Acres of these edifices, the tinge of grime declaring the relative

> dates of their erection; millions of tons of brute brick and mortar, crushing the spirit as you gaze.
>
> (p. 274)

This is not Benjamin's Berlin courtyard of his childhood 'where the city opened itself to the child' but the prison house of modernity closed in upon itself, though both are full of intimate, physical possibilities. These possibilities are exciting sexual ones for Benjamin, but distressing disease-laden ones for Gissing for 'the air was poisoned with the odour of an unclean crowd' (p. 274). Gissing subscribed to the miasmatic theory of disease commonplace at the time in which bad air (literally 'malaria') caused this and other diseases (see Giblett, 1996).

City conditions

Engels subscribed to the miasmatic theory too. In his 1892 'Preface to the English Edition' of his *Condition of the Working-Class in England* Engels (1987, p. 42) quotes an article of his own in which he had said that the East End of London in 1845 was 'an everspreading pool of stagnant misery and desolation.' In short, it was an industrial quaking zone of misery and desolation. He is pleased to report 40 years later that 'that immense haunt of misery is no longer the stagnant pool it was six years ago' (p. 45). Misery has not ceased but at least the East End is no longer the stagnant pool it was – due in his mind to unionism. Yet the 'great towns' of industrial capitalism are stagnant pools not merely metaphorically and morally but also literally and materially as 'the streets are generally unpaved, rough, dirty, filled with vegetable and animal refuse, without sewers or gutters, but supplied with stagnant pools instead' (p. 71). These stagnant pools in cities are worlds away from Thoreau's (1962, *VII*, p. 304) 'stagnant ponds or pools in woods floored with leaves.' From Engels' (1987, p. 71) 'filthy streets,' as would be expected, 'a horrible smell' arises. The Aire River of Leeds, 'thick, black, and foul, smelling of all possible refuse,' 'engenders miasmatic vapours' (p. 81) according to a journal reporting the sanitary conditions of working people in cities from which Engels quotes. Similarly, in Manchester:

> At the bottom flows, or rather stagnates, the Irk, a narrow, coal-black, foul-smelling stream, full of debris and refuse, which it deposits on the lower right bank. In dry weather, a long string of the most disgusting blackish-green slime pools are left standing on the bank, from the depths of which bubbles of miasmatic gas constantly arise and give forth a stench unendurable.
>
> (p. 89)

Hardly surprisingly Engels concludes that Manchester is 'this hell upon earth' which 'arouses horror' (p. 92).

Miasmatic gas is often described in short, and in a word, as effluvia that arise from 'masses of refuse, offal and sickening filth [that] lie among standing pools in all directions' with the result that 'the atmosphere is poisoned by the effluvia from these' (p. 98). Although Engels does not belabour the miasmatic theory of disease, he does not demur from repeatedly quoting the journal reporting the sanitary conditions of working people in cities to the effect that in such streets 'a mass of dried filth and foul vapours are created, which not only offend the [senses of] sight and smell, but endanger the health of the inhabitants in the highest degree' (p. 78). Later he more explicitly argues:

> The filth and stagnant pools of the working people's quarters in the great cities have, therefore, the worst effect upon the public health, because they produce precisely those gases which engender disease; so, too, the exhalations from contaminated streams.
>
> (pp. 128–129)

Certainly, as Engels says, 'life in large cities is, in itself, injurious to health,' but whether this can be attributed to what he calls 'gases decidedly injurious to health' (p. 128) is another question and more a matter of nineteenth-century misdiagnosis. Nevertheless, the city is a place of death for many nineteenth-century writers. For Thomson (1993, p. 29), 'The City is of Night; perchance of death.' Following in a similar vein Kipling (accessed online) is in no doubt in his short story 'The City of Dreadful Night,' published in 1891, that 'the city was of Death as well as Night.'

The stagnant pool in the city made with human hands is a deadly place. The city often had its beginnings in the stagnant pools of swamps not made with human hands. Many modern cities, or areas in them like the East End of London, for Davis (2006, p. 82) 'the Victorian world's greatest slum,' were built on reclaimed marshes, or drained swamps (see Giblett, 2016, chapter 4). For Neuwirth (2006, p. 179), 'All cities start in mud.' Cities end in man-made, dried-up crud. Mud for Thoreau (1962, *V*, p. 499) writing in his journal in 1853 is 'Nature's womb.' All cities start in nature's womb of mud and end up in culture's tomb of crud. Crud is culture's tomb. The city's end is in crud. The city is born from mud and dies in crud; it is born from nature's womb and dies in culture's tomb. The city made of mud-bricks or concrete, a mixture of the three elements of water, earth and air, or of clay-bricks baked in fire, a mixture of four elements, ends up in the elements falling apart and returning to dust. Yet cultural crud is often figured as natural mud. It is as if the filled or drained swamp on which the city was built returns in the fascination of the dark underside of the stagnant pools of the city, as if the spatial, geographic and historical repressed of the city returns in metaphors of the city. James Thomson (1993, p. 29) describes in 'The City of Dreadful Night,' his meditation on the modern city (obviously based on London), how:

> A river girds the city west and south,
> The main north channel of a broad lagoon,

Regurging with the salt tides from the mouth;
  Waste marshes shine and glisten in the moon
For leagues, then moorland black, then stony ridges;
Great piers and causeways, many noble bridges,
  Connect the town and islet suburbs strewn.

The river is a monster that girdles the city and regurgitates the black waters of waste wetlands.

London on London

The city makes a mark on the earth – in space – and marks a place in history – in time. Its dwellers, its citizens, make a mark on its streets. When Jack London (2001, p. 39) was in London in 1902 he remarked repeatedly on 'the slimy, spittle-drenched sidewalk.' He contrasts 'the solid walls of bricks' with 'the slimy pavements, and the screaming streets' (p. 4), the solid and vertical contrasted with the becoming liquid and horizontal. In this spatial poetics, if not metaphysics, the solid and vertical is valorized over the slimy and horizontal.

London had just witnessed:

> a market [where] tottery old men and women were searching in the garbage thrown in the mud for rotten potatoes, beans, and vegetables, while little children clustered like flies around a festering mass of fruit, thrusting their shoulders into the liquid corruption, and drawing forth morsels but partially decayed which they devoured on the spot.
>
> (pp. 3–4)

This urban, industrial swamp of decomposing and stinking vegetable matter is for London 'that wilderness' (p. 4). Similarly, for Morrison (1996, p. 10), 'The whole East End was a wilderness of slums.' Perhaps it was not unlike the swamp of the wilderness itself with their similar smells to the undiscerning nose to which all bad smells are bad in the same way. 'Filth and noisomeness' are coupled together for Mayhew (1985, pp. 109 and 177) too; bad matter and bad smells go together. Dirt as matter out of place is also noisome as the bad smell of a place. Whether that place was a slum or a swamp did not make much difference.

The slums of London (in two senses) are a site of horror as Jack goes on to state:

> For the first time in my life the fear of the crowd smote me. It was like the fear of the sea, and the miserable multitudes, street upon street, seemed so many waves of a vast and malodorous sea, lapping about me and threatening to well up and over me.
>
> (London, 2001, p. 4)

This fear of engulfment by the liquid and horizontal is evinced not only by the sight, sound and touch of the crowd and the city, but also by its smells.

Yet London finds:

> When at last I made into the East End, I was gratified to find that the fear of the crowd no longer haunted me. I had become a part of it. The vast and malodorous sea had welled up and over me, or I had slipped into it, and there was nothing fearsome about it.
>
> (p. 7)

The crowd takes on some of the qualities of the urban swamp as a few pages later London describes:

> a woman of the finest grade of the English working-class, with numerous evidences of refinement, being slowly engulfed by that noisome and rotten tide of humanity which the powers that be are pouring eastward of London town.
>
> (pp. 13–14)

The 'strange, vagrant odours' (p. 122) are like the vagrants who frequent the vagrant space of the slum. The human tide is flotsam and jetsam. The festering mass of fruit is ultimately indistinguishable from 'the festering contents of slum, stews and ghetto.' This human sewage 'resemble[s] some vile spawn from underground' (p. 87). This human sewage is not spawned so much from under*ground* but from what London calls 'the under-world of London' (p. vii), the city, and the underworld of London, the writer, whose tropes – 'the dreams of speech' as Nabokov calls them – 'give birth to the denizens of the slum and are 'the royal road to the unconscious' as Freud said of dreams. London 'went down into the underworld' of his eponymous city 'with an attitude of mind which I may best liken to that of the explorer' (p. xiii). He penetrates the urban underworld of the slum swamp just as the explorer in the colonies descended into the underworld of the swamp (see Giblett, 1996). Just as the romance hero emerged with his manhood vindicated through trial and tribulation, so does London (the writer) emerge to write the book and preface it in Piedmont, California, literally at the foot of the mountain, a dry, elevated place thousands of miles away from, and thousands of feet above, the wet, low place of the urban underworld of the East End of London (the city). Like Dante, he descends into 'the inferno' (pp. 1 and 6) and 'infernal regions' (pp. 52 and 68) of 'hell on earth' (p. 68) and the inferno of London (p. 159) to ascend into the paradise of heaven on earth of God's mountains in California as John Muir called them.

Similarly, Arthur Morrison's (1996, p. 1) *A Child of the Jago* begins with a description of:

> the narrow street all the blacker for the lurid sky; for there was a fire in a farther part of Shoreditch, and the welkin was an infernal coppery glow.

> Below, the hot, heavy air lay a rank oppression on the contorted forms of those who made for sleep on the pavement; and in it, and through it all rose from the foul earth and the grimed walls a close, mingled stink – the odour of the Jago.

For Morrison, the Jago, 'for one hundred years the blackest pit in London, lay and festered,' like an open, infected wound inflicted on the body of the earth. In the centre of the blackest pit was Jago Court, 'the blackest hole in all that pit.' This is the centre of the inferno where Satan resides in Dante's Inferno with the denizens of the slum as Satan in this secular theology, a world where God is dead. The sublime is God in secular theology and slime is Satan.

Like Morrison, London, the urban and colonial explorer, the endo- and exo-colonist, the colony within at home and the colony without away from home, is apt to find both decay and decomposition wherever he looks and treads. Although London and Morrison did not find miasma and malaria (unlike his colonial counterparts or previous urban explorers), Morrison noted 'the stifling air' (p. 2) and London (2001, p. 23) discovered 'disease germs that fill the air of the East End' and 'the effluvia and vile exhalations of overcrowded and rotten life' (p. 126). He also found that 'rottenness is a slimy desecration of the sweetness and purity of nature' (p. 87). The denizens of 'city slime' (p. 113) are slimy creatures of slimy pavements, 'a fearful slime that quickened the pavement with life' (p. 151). Human life has taken on a vegetable, animal and liquid life of its own. Humans are not autochthonous creatures that spring from the earth, or swamp, but creatures that spring from the urban swamp, the slum – they are auto-ex-metropolis, self-generating from the city. For London (p. 28), 'A spawn of children cluttered the slimy pavement, for all the world like tadpoles just turned frogs on the bottom of a dry pond,' or more precisely on the bottom of a dried-up industrial urban wetwasteland. The categories of solid and liquid, animal and vegetable, and the elements of earth, water and air are all mixed up. 'The place swarmed with vermin' (p. 83), with 'swarming children' (p. 91) and with workers who swarm (p. 113). In Morrison's (1996, p. 1) *A Child of the Jago*, in Jago Court 'the human population swarmed in thousands.' Swarming creatures neither walk, nor swim, nor fly, but are hybrid creatures, or monsters, who do all three.

As swarming creatures, they are neither fish, flesh, nor fowl, and they fall under the Levitical interdiction of being an abomination. For London (2001, p. 152), 'They are a new species, a breed of city savages. . . . The slum is their jungle.' This is not a tropical rainforest where nature is green in leaf and branch, but an urban jungle where culture is red in tooth and claw. For London it would be 'far better to be a people of the wilderness and desert . . . than to be a people of the machine and the Abyss' (p. 153). Better to be creatures of the swamp and jungle than of that urban swamp and jungle called the slum; better to be creatures of the living waters of wetlands than to be minced to death in 'the Abyss [which] is literally a huge man-killing machine' (p. 23) just as the slimy trenches of World War I were later; better to be a living swamp creature than to be 'the living deaths' (p. 152) at 'the bottom of the Abyss' (p. 152); and better to be in the jungles of 'Darkest Africa' (p. 1) than in 'the jungle of empire' (p. 149).

The landscape of urban slums is not only prescient of the landscape of trench warfare, but also the product of class warfare perpetrated by the military-industrial ruling class. For London, 'Here, in the heart of peace, is where the blood is being shed' (p. 135). 'In London the slaughter of the innocents goes on on a scale more stupendous than any before in the history of the world' (p. 150), including Herod's in the wake of the birth of Jesus (p. 135). This slaughter is perpetrated by 'the men of England, masters of destruction, engineers of death . . . men of steel . . . war lords and world harnessers' who have 'mastered matter and solved the secrets of the stars' (p. 76). London also maintained that if the masses of the slums attempted revolt they would perish 'before the rapid-fire guns and the modern machinery of warfare' (p. 123). Rather than revolting, they ended up as cannon fodder in the 'meat-mincer' of the fields of Flanders during World War I (see Giblett, 2009, chapter 4).

The denizens of the East End are monstrous like the city they inhabit (hardly live in). The city takes on a life of its own as it is what London calls 'the monster city' (p. 14) with its own aberrant and peculiar anatomy. Night in the city is not merely the time of sunlessness but 'the black night of London [which] settles down in a greasy pall' (p. 14). London for James Thomson is 'the City of Dreadful Night,' but not for Jack London. The East End is 'often called,' according to London, 'the City of Dreadful Monotony' (p. 113). Just as wilderness swamps were often called dreadful and monotonous, so the urban swamp is invoked in similar terms, though for London 'the East End does merit a worse title. It should be called 'The City of Degradation'' (p. 114). It takes one lower into the depths and bowels, not of the good earth and wilderness swamp, but of the evil city and city swamp, 'the perilous depths of the East End,' as Morrison (1996, p. 10) puts it. In this body politics of the city the park is not for London (2001, p. 31) 'a lung of London,' which it should be in accordance with the dictates of the desirable Victorian body-politic, but 'an abscess, a great putrescent sore' on the bottom or backside of the body-politic that should be treated and the patient cared for, not lanced in the gentrification of slum clearance.

The delta is the cloaca of the body terrestrial. St Petersburg is another famous modern city built on the reclaimed wetland of a delta. For W. G. Sebald (2002, p. 48):

> marooned on the Neva's marsh delta
> St. Petersburg under the fortress,
> the new Russian capital,
> uncanny to a stranger,
> no more than a chaos erupting,
> buildings that began to subside
> as soon as erected, and nowhere
> a vista quite straight. The streets

and squares laid out according
to the Golden Section, jetty walls and bridges,
alignments, façades and rows of windows –
these only slowly come towards us
out of the future's resounding emptiness,
so as to bring the plan of eternity into the city
born of the terror of the vastness of space

The city not only has a dark underside but it was also built on repressing that other side in order to keep at bay not only the sublime terror of the vastness of space outside its walls, gates, suburbs, etc., but also the slimy horror of places inside and beneath and before it, under its streets, houses, buildings, etc. Sebald has read Thomson (p. 64) for the latter refers to 'vast wastes of horror-haunted time.' The city of dreadful night is a bulwark, not only against the terror of space, but also against the horror of time. The city not only marks out a grid of allotments on the earth, but also marks a place in history. The quaking zone of the swamp, by contrast, mediates between time and space, place and history in the cyclical now-time of birth-death-birth.

Like St Petersburg, Perth, Western Australia, was founded in a wetland and marked out across its wetlands. It is not only 'A City of Wetlands' but also a 'grid-plan' town, both in its early colonial and recent neo-colonial phases (see Giblett, 1996):

Perth, uncanny city
of the concrete tunnel
that snakes its way
through the underground watercourse
of buried wetlands
to disgorge its cargo
of cars and trucks
on the rubbish-strewn
banks of the black swan river
close by where
the natural outflow of the wetlands
used to be
through Claise Brook
now a yuppified lagoon
shaped like it used to be
according to old maps
but contained now in limestone walls
surrounded and overseen
by a 'livable neighbourhood'
of a soon-to-be yuppie slum
of inner urban renewal
and redevelopment of the old slum

the walls of the eastern end of the tunnel
are painted with the Waugyl
to commemorate the water-being
that once and still does (sort of)
wend its way through the buried wetlands
its life-giving/death-dealing waters
replaced by traffic flows
through the 'polly pipe'
another token commemoration
of indigenous contribution

the tunnel displaced
the underground flows,
the wetlands drained above ground
were drained underground too
the land shrank, the houses cracked
as every engineer should have known they would
old maps showed the wetlands that were there
history showed when you drain the wetlands,
the land shrinks
like a sponge with the water wrung out of it

the buried history returned with a vengeance
the tunnel is a dam
that blocks the underground flows
the residents of Northbridge
wake after a deluge
and wander into their kids' rooms
and find themselves
ankle-deep in water

the tunnel is damming the flows
that still run deep
the history is buried but not lost
the tunnel floats on these flows
the tunnel is cracking too

The way that various writers describe and figure the city plays out their fears.

James Thomson figured the dark underside of the late nineteenth-century city as dreadful night; Engels as a haunt of misery. Antoine Picon (2000) recently characterized this side of the late twentieth-/early twenty-first-century city as an 'anxious landscape.' Picon contrasts Manhattan as 'a magic city made of crystal' and as 'the celestial Jerusalem' with 'the hell, or purgatory' of Newark with its 'creeping swamps' where everything is 'rusted out' and 'irreparably polluted yet somehow endowed with a strange beauty.' In Yeatsian terms, a terrible beauty is born in the rusted city. The central business district with its sublime skyscrapers is heaven compared to the slimy hell of industrial swamps.

Picon is working in the tradition of Jerrold's and Doré's (2005, p. 11) depiction of nineteenth-century London, in particular of 'the dead shore' of the Thames River with 'the muddy, melancholy bank' with 'rust upon everything.' As the marsh has traditionally been associated with melancholy, Jerrold's and Doré's pilgrimage could have referred to the muddy, melancholy bank as 'a slough of despond' to make an explicit connection with John Bunyan's *Pilgrim's Progress* (see chapter 6). Picon does not use such archaic language, but the implication or impression is that the Newark industrial swamps are depressing, though strangely beautiful, places.

Similarly when Jerrold and Doré (2005, p. xxviii) went on their 'travel[s] about London in search of the picturesque' they 'discovered that it abounded . . . in picturesque scenes.' The picturesque is not an object, nor a feature of the object, but a point of view, or a modality that sets up a relation between the subject and object, the viewer and the scene and seen. If one goes in search of the picturesque, one will invariably find it because one carries the picturesque point of view with one. For Jerrold (2005, pp. 4 and 5):

> The lesson which Doré's pictorial renderings of our mercantile centre will teach . . . is that London, artistically regarded, is not, as the shallow have said so often, an ugly place, given up body and soul to money-grubbing.

Yet there are limits to the picturesque as Jerrold (2005, p. 15) concedes that in the poverty of the poor 'there is nothing picturesque.' Poverty cannot be depicted in picturesque terms because the picturesque is an impersonal point of view from a distance whereas poverty is personal (see Doré's illustration no. 119, 'Wentworth St, Whitechapel,' (Jerrold and Doré, 2005, p. 145)). The poverty of the poor is conveyed much better in Jerrold's words used by Coolidge (1994, p. 53) as a caption to this illustration, than in Doré's drawing: 'beat-up alleys with pools of water. Here, tattered young women . . . look at you sullenly, fiercely, what suffering, what hunger can be read in their meager bodies, hungry since the day they were weaned.' Jerrold does not reflect on the inability of the picturesque to convey the poverty of the poor and the ability of language to do so. The picturesque is also unable to convey the poverty of impoverished landscapes. Rusting industrial ruins and other wastelands have become an object of contemporary fascination, and fashion. See, for example, the photographic work of Edward Burtynsky, a kind of industrial chic (see Giblett, 2009, chapter 9).

The dark sides of both the late nineteenth-century and late twentieth-/early twenty-first-century cities are invoked in terms of the swampy and uncanny and testify to a fascination with, and horror of, the urban nether world and the (in this case, artificial) quaking zone known otherwise and simply as 'the slum' or, as Neuwirth (2006, p. 241) prefers to avoid the pejorative connotations, 'squatter communities.' The slum in the early twentieth-first century is the habitat for one billion people. For Davis (2006, p. 19), 'Instead of cities of light soaring toward heaven, much of the twenty-first-century urban world squats in squalor, surrounded by pollution, excrement and decay.' Squats, in other words, in a man-made swamp, and not only squats physically on its haunches, but also illegally on land its residents do not own (Neuwirth, 2006, p. 12). Squatting today for Davis (2006, p. 39) 'continues primarily in low-value urban land, usually in hazardous or extremely marginal locations such as floodplains, hillsides, swamps or contaminated brownfields.' The slum for Davis (2006, p. 22) was 'first and above all envisioned as a place where an incorrigible and feral social 'residuum' rots in immoral and often riotous splendour.' Vegetable decomposition figures moral decay and anarchic display. The slum is the arsehole or cloaca of the body-politic, a festering wound inflicted on the body of the earth. Whether or not it can, or how it should, be treated is another matter.

For Davis (2006, p. 95) 'urban inequality in the Third World is visible even from space: satellite reconnaissance of Nairobi reveals that more than half of the population lives on just 18 percent of the city area.' This area includes Kibera, 'Africa's largest mud hut metropolis' (Neuwirth, 2006, p. 70), or slum. Satellite reconnaissance of Nairobi also reveals nearby Nairobi National Park, the world's largest National Park within the municipal bounds of a city. Two iconic and interdependent features of modernity sit cheek-by-jowl with each other: nationalism and landscape preservation and display in the national park side-by-side with the underbelly of urbanity in the slum; the national park with vast open areas populated sparsely by rangers and tourists with the slum crowded and over-built with rusting iron shanties and muddy, sewery paths; and the national park created by the dispossession and removal of its native owners and inhabitants into the slum.

The British for Davis (2006, p. 52) have the dubious distinction of being 'the greatest slum builders of all time.' They designed and built railways (or at least their labourers did), and the railways, in turn, for Coolidge (1994, p. 32) were 'the greatest creators of London slums.' British railways, at home and abroad, created the interstices between railway lines that were filled by slums. The communication and transportation technology of the railway excommunicated slum dwellers between and beside its lines. Nairobi, for example, came into being in 1899 when the British wanted to span East Africa with a railway and used a small Masai settlement at the confluence of several small rivers as the staging point for the construction of the railway (Neuwirth, 2006, p. 91).

Railways created the slum during the period from the nineteenth century to the early twentieth century. Road transport and satellite communication are the greatest creators of planetary slums today that excommunicate slum dwellers between its highways and vectors. Yet the sanitary conditions are pretty much the same. Many slum dwellers today live in what Davis (2006, 144) calls 'the sanitary equivalent of the mud hell of World War I trench warfare.' The slum is the product of class war fought against the lower classes. Intra-urban class warfare produced a landscape very similar to that produced by international armed warfare. Thirty years before Davis, Fussell ((2000, p. 149) had noted the converse, 'the similarity of the trench scene to modern, urban, industrial squalor.' It is hardly surprising that both places end up looking and being the same as both were the product of class warfare fought against the earth, against both the lower earthly and class strata (for the 'mud hell' of World War I trench warfare see Giblett, 2009, chapter 4).

The city founded in war, or at least of the preparations for it as Virilio put it (cited in Giblett, 1996, p. 56), becomes the landscape of warfare, whether it is the landscape of World War I trench warfare or the landscape of World War II aerial warfare. 'The urban institution of war,' as Mumford (1961, p. 42) puts it, makes the city into the landscape of warfare. This landscape is either the landscape of class and trench warfare as in WWI or the landscape of international and aerial warfare as in WWII. What Mumford (1961, p. 53) following Geddes calls 'a living urban core, the polis, ends in a common graveyard of dust and bones, a Necropolis, or city of the dead: fire-scorched ruins, shattered buildings, . . . heaps of meaningless refuse' – in a word crud, as in the aftermath of the fire-bombing of Dresden, Hamburg, Wurzburg (see Giblett, 2009, chapter 5).

Hans Erich Nossack (2004, pp. 1–2) witnessed the destruction of Hamburg as a spectator from outside the city. Hamburg for Nossack (2004, p. 32) was 'the first big city to be annihilated' in the WWII. He bore witness to the destruction in his book called *The End*. The title refers not only to the end of the city as he knew it, but also the end of the world in time and space in apocalyptic times and tones. What he calls 'the ruin of Hamburg' (Nossack, 2004, p. 6) is also variously described in apocalyptic terms as 'the abyss' (Nossack, 2004, p. 6), 'the end' (Nossack, 2004, p. 8) and 'the netherworld' (Nossack, 2004, p. 11), terms which have also been applied to the slummy underside of other cities, such as London (as we have seen). Not merely 'anti-city,' the destruction of Hamburg by fire also pitted nature against nature for 'even nature had risen up in hatred against herself' (Nossack, 2004, p. 11). The military strategy of modern aerial and naval warfare is anti-nature as two forces of nature (fire, storm) are both harnessed by militaristic men and unleashed against nature and the earth. All wars are fought against nature and the earth and produce a war-torn landscape of destruction and ruination. The result for Nossack is 'the raging of the world against itself' and 'the earth writhed in agony' (Nossack, 2004, p. 15). These are the death throes of 'raging destruction' (Nossack, 2004, p. 13) inflicted by the world and the earth upon itself by a world- and earth-hating military whose end (in two senses) is 'the ruined world' (Nossack, 2004, p. 13). 'The horror' of 'the disaster' (Nossack, 2004, p. 16) was that it was both natural and cultural, the forces of nature directed against both nature and culture by natural and cultural human beings against other natural and cultural human beings.

Part of the horror for Nossack was the fact that the refugees fleeing from the city 'brought with them an uncanny silence. No one dared question these mute figures' (Nossack, 2004, p. 17). They had ceased to be human, to be natural and cultural beings, because they were deprived of speech and had become speechless automatons that mimicked and mocked human behaviour in an uncanny manner. The fact that this disastrous destruction of Hamburg was 'the end' for its residents of 1943 meant 'we no longer have a past' (Nossack, 2004, p. 23), only a present that was unbearable and an unthinkable future torn away from any connection with the past. Perhaps they did not even have a present and future as 'we no longer had any time at all, we were outside of time.' In other words, they transcended time into eternity (Nossack, 2004, p. 40). Space and time ceased to exist as they had been known to exist in the past.

Not only was the spatial landscape transformed, but also the temporal timescape as well: 'The infinite behind man wafted unhindered in the endlessness before him and hallowed his countenance for the passage of what is beyond time' (Nossack, 2004, p. 29). Space and time are collapsed together into the landscape of eternity and infinity, into the sublime where space and time are collapsed. By mixing fire and air and sublimating solid matter into gaseous flame and smoke, the fire-bombing of cities in Second World War was sublime (see Giblett, 2009, chapter 5). The less, or more, than human figures who people this landscape are not only mute automatons but also icons with eyes 'grown larger and transparent' (Nossack, 2004, p. 29) through witnessing the horrors of fiery destruction of their city and the (its, their) past.

City as body

The city has been figured in a variety of ways (as we have seen in this chapter). These have ranged from abyss to jungle, nether world to hell, sublime heights to swampy depths, and so on. The ways in which the city is figured say much about the politics of the period in which the figuration was made. Yet these historical shifts say a lot not only about those periods (and so are of historical interest) but also about the city, how it has been thought from the past into the present and how it will be thought in the future (and so are of contemporary relevance). In concluding this chapter I look at one of the central ways in which the city has been figured, the city as body, and variations on this theme. I also relate this trope to a much older one of the earth as body and the body as earth and argue for a rapprochement between them in such a way that the city can become more 'sustainable,' in other words, live in a more psycho- and bio-symbiosis in its bioregion and with the living earth.

I propose remythifying the city as a body, not only with parks as lungs as in the nineteenth-century cliché of the city, but also with its twentieth-century skyscrapers as the head and brain; the water supply and sewerage systems as the oesophagus and intestines; the rivers as arteries; the wetlands as kidneys, liver and placenta; the 'mouth' and the estuary of the river as the anus and bowels; slums as a sore to be treated and cared for, not bled or lanced in the gentrification of slum clearance; and so on. Rather than figuring slums as swamps in pejorative terms as places of disease and horror, I propose remythifying them as artificial wetlands whose livability for their residents would be improved by improving their ecological functionality as kidneys, liver and placenta, as places of hope and new life.

Thinking of the earth as body and the city as part of that body with a range of organs and processes can enable urban earth dwellers to live in a more symbiotic relationship with the earth. Nabokov said tropes are the dreams of speech, and Freud said dreams are the royal road to the unconscious. Tropes lead to the unconscious of culture and the city. Detheologizing and remythifying the pejorative tropes of the city and the earth (as this book tries to do) and figuring both otherwise in more symbiotic terms may enable city and country dwellers to respect and care for the earth more than we do at present.

References

Benjamin, W. 1979. *One-Way Street and Other Writings*. E. Jephcott and K. Shorter, trans. London: NLB.

Benjamin, W. 1999. *Selected Writings, Volume 2: 1927–1934*. R. Livingstone and others, trans. M. Jennings, H. Eiland and G. Smith, eds. Cambridge, MA: Belknap Press of Harvard University Press.

Coolidge, J. 1994. *Gustave Doré's London: A Study of the City in the Age of Confidence 1848–1873*. Dublin, NH: William L. Bauhan.

Davis, M. 2006. *Planet of Slums*. London: Verso.

Douglas, M. 1966. *Purity and Danger: An Analysis of the Concepts of Pollution and Taboo*. London: Routledge.

Engels, F. 1987. *The Condition of the Working Class in England*. V. Kiernan, ed. London: Penguin. First published 1845.

Freud, S. 2003. *The Uncanny*. D. Mclintock, trans. London: Penguin. First published 1919.

Fussell, P. 2000. *The Great War and Modern Memory*. New York: Oxford University Press. First published 1975.

Giblett, R. 1996. *Postmodern Wetlands: Culture History Ecology*. Edinburgh: Edinburgh University Press.

Giblett, R. 2009. *Landscapes of Culture and Nature*. Basingstoke: Palgrave Macmillan.

Giblett, R. 2016. *Cities and Wetlands: The Return of the Repressed in Nature and Culture*. London: Bloomsbury.

Giblett, R. 2018. *Environmental Humanities and Theologies: Ecoculture, Literature and the Bible*. London: Routledge.

Gissing, G. 1992. *The Nether World*. Oxford: Oxford University Press. First published 1889.

Jerrold, B. and G. Doré. 2005. *London: A Pilgrimage*. London: Anthem Press. First published 1872

London, J. 2001. *The People of the Abyss*. London: Pluto Press. First published 1903.

Mayhew, H. 1985. *London Labour and the London Poor*. London: Penguin. First published 1861–62.

Morrison, A. 1996. *A Child of the Jago*. Chicago: Academy. First published 1896.

Mumford, L. 1961. *The City in History: Its Origins, Its Transformations, and Its Prospects*. New York: Harcourt.

Neuwirth, R. 2006. *Shadow Cities: A Billion Squatters, a New Urban World*. New York: Routledge.

Nossack, H. 2004. *The End: Hamburg 1943*. J. Agee, trans. Chicago: University of Chicago Press.

Picon, A. 2000. *Anxious Landscapes: From the Ruin to the Rust*. Available online at: www.gsd.harvard.edu/people/faculty/picon/texts.html. Accessed 29 September 2006.

Sebald, W. 2002. *After Nature*. M. Hamburger, trans. New York: The Modern Library.

Solnit, R. 2000. *Wanderlust: A History of Walking*. London: Penguin.

Thomson, J. 1993. *The City of Dreadful Night*. Edinburgh: Canongate Classics. First published 1880.

Thoreau, H. 1962. *The Journal of Henry D. Thoreau, Volumes I- XIV*. B. Torrey and F. Allen, eds. New York: Dover.

4 The uncanniness of Schelling's uncanny

At a time and for a world in which some sectors of the monotheisms are in aggressive, and sometimes violent, contention against those of other monotheisms, and even against other sectors within their own monotheism, Friedrich Schelling's discussion of the uncanny enables rethinking the relationship between mythology and monotheistic and polytheistic religions in a culturally and politically liberatory and progressive way. The monotheistic sectors generally claim special and privileged access to the truth by way of divine revelation in their scriptures and/or via their priests, and the more fundamentalist ones don't acknowledge their roots in mythology or, if they do, in polytheism. By equating mythology with polytheism, Schelling prioritizes polytheism over monotheism, whereas polytheism is anathema for the Judaic and Christian monotheisms as proclaimed in the first interdiction of the Ten Commandments, 'You shall have no other gods before me' (*Exodus* 20: 3). Schellingian polytheism deconstructs this commandment by countermanding to the monotheistic God in dialogic reciprocity, 'There are many other gods before you.'

These aspects of Schelling's thinking about the uncanny are absent from Sigmund Freud's classic essay on the uncanny. Freud's engagement with Schelling's uncanny is sketchy, superficial and peremptory to say the least. Freud (2003, p. 132) cites Schelling's discussion of the uncanny in his essay on the uncanny but does not give a precise reference. Schelling's uncanny is relatively hidden and secret in Freud's essay and only occasionally comes to light or out into the open in later discussions (like the uncanny itself). Schelling's uncanny is uncanny in Freud's essay. Similarly, some Anglophone Schellingians mention a source for Schelling's definitions of the uncanny, but these references are often as uncanny as Freud's reference in his essay to Schelling's uncanny and to the story that he recounts that he had found in *The Strand* magazine about a table with carved crocodiles and that he thought exemplifies the uncanny (Freud, 2003, p. 151; Moberly, 1917, 1991; Giblett, 2009, chapter 2, 2018, chapter 4).

Schelling's uncanny is not only uncannily present/absent in Freud's essay, but also completely absent from some studies by Anglophone writers on Schelling, mythology and the unconscious. For instance, S. J. McGrath's tantalizingly entitled *The Dark Ground of the Spirit: Schelling and the Unconscious*, in which one might expect to find a discussion of the uncanny given the title and subtitle, does not mention or discuss the uncanny at all (McGrath, 2012). Similarly, Victor Hayes' *Schelling's Philosophy of Mythology and Revelation*, a translated 'reduction' of and commentary on Schelling's three lecture series on these topics, in the second series of which he defines and discusses the uncanny, is in the same boat as McGrath with no mention of it (Hayes, 1995). By contrast, some writers more interested primarily in Freud and his concepts of the unconscious and/or the uncanny (rather than in Schelling and his discussion of the unconscious and/or the uncanny) address the uncanny in Schelling's work.

This chapter argues that Schelling's discussion of the uncanny taps into a timely and rich vein of thinking about the divine, darkness, light, aura, religion, mythology and the monstrous overlooked by Freud that enables rethinking the inter-relationship between all of them in less pejorative terms than Christian theology does, including ecotheology, and in more environmentally and animal friendly terms (as ecocultural studies does; see Giblett, 2018, chapter 4). As well as locating the primary source of Schelling's definition of the uncanny cited by Freud, the chapter interweaves this bibliographic solution with a discussion of mythology and polytheism in Schelling's work and in the secondary literature. The chapter is variously: a bibliographic detective story; a contribution to the burgeoning literature on the uncanny; a contemporary political and theological intervention into monotheistic contention; and a work of environmental philosophy and activism by rethinking religion through mythology and poetics as a way of bringing being and living with the earth into closer affinity with nature.

Schelling's uncanny and Freudian psychoanalysis

In his essay on the uncanny first published in 1919, Freud (2003, p. 132) cites Schelling's definition: 'Uncanny is what one calls everything that was meant to remain secret and hidden and has come out into the open.' Rather than translating '*hervorgetreten*' (Freud, 1955, p. 256)[1] as 'come out into the open,' the *Standard Edition* of Freud's works translates the German as 'come to light' (Freud, 1955, p. 224). '*Hervorgetreten*' is usually translated as 'emerged,' to which the later 2003 Penguin translation seems closer than the older *Standard Edition* translation, which implies enlightenment both as an individual process and philosophical moment and movement. Terry Castle (1995, p. 7) bases her study of the invention of the uncanny as the counter to enlightenment in eighteenth-century culture on the flimsy foundation of this dubious translation in the *Standard Edition*.

Schelling emphasizes more the religious and mythological aspects of the uncanny than the philosophical and psychological ones. Freud does not cite Schelling directly but cites his definition from a secondary text in the form of Daniel Sanders' *Wörterbuch der Deutschen Sprache* (*Dictionary of the German Language*), published in 1860, that, in turn, cites 'Schelling 2.2, 649' (more of this cryptic reference later, which the translation in the *Standard Edition* omits). The precise source of Schelling's definition remains hidden in Freud's essay and does not come out into the open. Freud later appropriates Schelling's definition for his own definition of the uncanny in terms of psychological repression, whereas

Schelling's definition emphasizes polytheistic mythology. Schelling's philosophy is not about repression, but release into the possibility of a new mythology.

The hidden and secret aspects of the uncanny relate particularly to Schelling's second mention of the uncanny that Freud (2003, p. 132) cites from Sanders: 'To veil the divine and surround it with an aura of the uncanny.' Schelling alludes here to the holy of holies in the Jewish temple that is veiled. Once a year the high priest is permitted to part the veil and enter the holy of holies to commune with the divine. As a result, the divine is veiled with 'an aura of the uncanny.' Schelling is using this example as a metaphor for the absolute, or the 'thing in itself.' Freud does not discuss these aspects of Schelling as cited by Sanders and they remain hidden and do not come out into the open in Freud's essay with this one exception. Aura and the uncanny are qualities that Walter Benjamin (2015, pp. 64, 69 and 72) found in early photographs, including in a photograph of Schelling taken in *c.*1850. Benjamin seems to have been unaware of Schelling's writings on aura and the uncanny, though he would have probably been aware of Schelling's post-Kantian idealism. As with Freud, Schelling's uncanny is uncannily present/absent in Benjamin's work.

Freud (2003, p. 123) generally secularizes, internalizes and psychologizes the uncanny and divests it of divine connotations in his own definitions of the uncanny at the beginning of his essay as belonging 'to the realm of the frightening, of what evokes fear and dread.' Of course, the divine can be frightening and evoke fear and dread, but Freud does not say so. On the following page, Freud more explicitly defines the uncanny as 'that species of the frightening that goes back to what was once well-known and had long been familiar' (Freud, 2003, p. 124). The divine may go back to what was once well known and had long been familiar, but at a cultural and historical level before the fall from grace, rather than at an individual one, especially before the return to grace. The divine for Schelling remains veiled and surrounded with an aura of the uncanny. The divine is precisely that which is *not* well known and familiar. The divine is mediated by revelation. Revelation for Schelling (2008, p. 100; emphasis in original) is 'something mediated through earlier processes, never . . . something *immediate*, first, original.' By contrast and implication, mythology is immediate, first, original.

Freud implicitly builds into his definition of the uncanny the tried and true concept and activity of repression. Freud (2003, p. 134) later explicitly evokes repression by using Schelling's definition, though for Schelling the uncanny is about the absolute shining imperfectly through the real, about hope, not repression. Sixteen pages later in the 2003 Penguin translation of Freud's essay after first citing Schelling's definition cited by Sanders, Freud (2003, p. 148) states that the uncanny is:

> actually nothing new and strange, but something that was long familiar to the psyche and was estranged from it only through being repressed. The link with repression now illuminates Schelling's definition of the uncanny as 'something that should have remained hidden and has come into the open'.

The uncanny Schellingian uncanny returns in Freud's uncanny, but Freud's other citation of Schelling's uncanny in terms of the veiled and auratic divine remains buried and hidden in Freud's essay in the form of Sanders' citation of Schelling and the cryptic reference, and never comes out into the open again. Freud's appropriation of Schelling's uncanny serves a repressive theory of the ego and blocks the possibility of thinking of *unheimich* as hope, rather than repression.[2]

Schelling gets short shrift from other writers on the uncanny too, more so than he does from Freud. In her recent 'genealogy of the uncanny,' Masschelein (2011, pp. 2 and 47) acknowledges that 'crucial for the uncanny is the concept of repression and the return of the repressed,' as it was for Freud referring to Schelling. Yet she does not acknowledge or cite, as Freud does, Schelling's definition of the uncanny. Schelling is largely written out of her study as signalled in the subtitle of *The Freudian Uncanny in Late Twentieth-Century Theory*. Even his surname is uncannily located out of alphabetical order in the index to her book after the entry headed 'Scientific' (Masschelein, 2011, p. 227).

Freud for Masschelein is the Promethean 'culture hero' who stole the concept of the uncanny from the gods and brought it down to mere mortals, whereas in fact he was the borrower of the concept from others, including Schelling and his writings on the gods. Masschelein credits Freud with raising:

> the phenomenon and the word '*unheimlich*' to the status of a concept in the foundational essay 'The Uncanny.' [. . .] Freud's essay [. . .] remains the primary focus of attraction in the continuing fascination with the uncanny in culture and theory alike. [. . .] This brings us to the central thesis of this book, namely that *the Freudian uncanny is a late-twentieth century theoretical concept.*
>
> (Masschelein, 2011, pp. 3–4; her emphases)

Masschelein goes on to mention Schelling in passing in this context, but her genealogy of the uncanny does not extend beyond Freud and his essay of the early twentieth century into discussing Schelling's work on the uncanny in the nineteenth century in which the phenomenon and the word '*unheimlich*' had already been raised to the status of a concept. If not, how else could Sanders and Freud cite Schelling's definition and the latter use his definition in discussing repression? Freud has no paternal rights of primogeniture when it comes to the concept and word 'uncanny,' certainly not to the 'discovery and creation of the concept in 1919,' as Masschelein (2011, p. 4) also puts it, nor does Schelling for that matter who traces the genealogy of the phenomenon back to pre-history and pre-mythology. If Freud discovered the concept, he could not have created it.

Alix Strachey is the midwife at this strange, cross-century and cross-cultural birth as she first translated Freud's essay '*Das Unheimlich*' into English as 'The "Uncanny"' (and *unheimlich* as uncanny) for Freud's *Collected Papers* first published in 1925, which was then 'considerably modified' for her husband James' *Standard Edition of the Complete Psychological Works of Sigmund Freud.* An editor's note introduces Freud's essay and states that 'the first section of the present paper, with its lengthy quotation from a German dictionary, raises special difficulties for the translator. It is to be hoped that readers will not allow themselves to be discouraged by this preliminary obstacle, for the paper is full of interesting and important material, and travels far beyond merely linguistic topics' (Freud, 1955, p. 218). *Unheimlich* remains untranslated in the extract from Sanders in the *Standard Edition*, including the citation from Schelling (Freud, 1955, p. 224). A footnote for the first appearance of the translation in the body of the text states that 'the German word, translated throughout this paper by the English "uncanny," is "*unheimlich*," literally "unhomely." The English term is not, of course, an exact equivalent of the German one' (Freud, 1955, p. 219). No kidding!

Masschelein (2011, p. 3) asserts archly that 'a theory of the uncanny before the twentieth century can only resort to the occurrence of the word or to descriptions of the phenomenon in literary texts and artistic sources. . . . There was no theoretical or philosophical discourse [on the uncanny] before the twentieth century.' Yet Schelling developed a theory of the uncanny in the nineteenth century in his philosophical texts (as the present chapter will demonstrate). Masschelein does not discuss Schelling's work and test her assertion (let alone substantiate it). In an endnote Masschelein, like others (as we will see below), refers to Schelling's *Philosophy of Mythology* as the source of his definition without citing a precise reference for it in Schelling's oeuvre (Masschelein, p. 171 n.22). Schelling's uncanny is hidden in her study and never comes out into the open.

A theory of the uncanny during and before the twentieth century can resort (and does so in the present chapter) to Schelling's definition and discussions in his philosophical texts. The central thesis of the present chapter is that in the early twentieth-first century the Schellingian uncanny is a theoretical concept that deserves to become a focus of attraction in the continuing fascination with the uncanny in culture and theory, in politics and practices alike, especially in the contemporary context of contention between some sectors of the monotheisms and Schelling's prioritizing of polytheism not merely before, but also over monotheism.

Despite the important status of Freud's essay and the attraction it still enjoys, it has a marginal place in discussions of the Freudian canon and psychoanalysis. This is despite Masschelein claiming that 'at the end of the twentieth century, [Freud's] rather short treatise had outgrown "its marginal position in the Freudian canon" and is now a central text for Freudian aesthetics' (Masschelein, 2011, p. 3, citing David Ellison). The uncanny is arguably anti-aesthetic as it counters the aesthetics of the sublime, picturesque and beautiful (see Giblett, 1996, chapter 2). The uncanny assaults the sense of smell, whereas aesthetics addresses the senses of sight and hearing (as do the sublime, picturesque and beautiful). Freud's essay has also arguably not outgrown its marginal position in the Freudian canon, Freudian psychoanalysis and in biographies of Freud. Over the last half a century, Freud's essay on the uncanny has been, and still is, marginalized in the Freudian canon by theorists of Freudian psychoanalysis and by recent biographers of Freud (as we saw in a previous chapter).

Schelling's uncanny and Anglophone Schellingians

Unlike Freud, some other writers on Schelling ostensibly go to a primary source in Schelling's later work and writings for the source of his definition of the uncanny. For instance, Jason Wirth (2003, p. 150), a leader in Anglophone Schellingian studies, in his first book-length study of the philosopher, *Conspiracy of Life*, introduces Schelling's definition of the uncanny by saying that he gave it in 'the *Philosophy of Mythology* lectures.' However, like Masschelein, this book does not appear in Wirth's bibliography of Schelling's works (p. 267), nor does it exist in English as such (as Wirth, like Masschelein, seems to suggest or imply that it does by giving it an English title). In the body of the text Wirth then cites the following translation of Schelling's definition: 'All that which should have remained in secret [*im Geheimmis*], in concealment and latency, but which has nevertheless stepped forward.' Wirth (2003, p. 254, n.31) gives the following endnote here for this citation: 'I have used Edward Allen Beach's translation of this passage in his *The Potencies of the God(s): Schelling's Philosophy of Mythology*, (Albany: State University of New York, 1994), p. 228.' The source of Schelling's definition in Schelling's work remains hidden in Wirth's work and does not come to light in it, whereas it does in Beach's study when he translates and refers to Schelling (Beach, 1994, p. 289, n.72). Beach also briefly mentions that Schelling's discussion of the uncanny 'later struck Sigmund Freud as particularly acute' and refers to his essay (Beach, 1994, pp. 229 and 289, n.74.)

Anthony Vidler (1992, p. 14) in the introduction to his study of *The Architectural Uncanny* first cites Freud and his later elaboration of Schelling's definition without acknowledging Schelling. Vidler elides Freud's reference to 'Schelling's definition' by using ellipses. Schelling remains hidden until Vidler (1992, p. 26; his ellipses), like Wirth, later cites Sanders citing Schelling and uses the following translation of Schelling's definition: '*Unheimlich* is the name for everything that ought to have remained . . . secret and hidden but has come to light.' Vidler (1992, p. 230, n.21) at this point refers to Sanders' *Dictionary* in an endnote. Vidler (1992, p. 26) goes on to state in the body of the text that 'Schelling's felicitous "definition" of the uncanny had in fact been extracted by Sanders from the *Philosophie der Mythologie* of 1835, Schelling's late attempt to synthesize the history of religion with the anthropology of cults.' Unlike Freud, Vidler also explicitly goes on to refer to Schelling's work when he cites Schelling's definition again on the following page in a slightly different (hitherto unpublished) translation by Eric Randolf Miller: 'All things are called uncanny which have remained secret, hidden, latent, but which have come to light.'

Vidler (1992, p. 27) also cites the context for Schelling's definition of the uncanny in which Schelling refers to 'the dark and obscure power of that uncanny principle which dominated earlier religions.' Darkness for Schelling has no pejorative overtones (as we will see later). He is not valorizing later religions over earlier ones, enlightened ones over benighted ones, light over darkness. The uncanny is the process of the dark coming into the light, the obscure coming into the open. As the source of Schelling's definition of the uncanny and enunciation of the uncanny principle, Vidler (1992, pp. 27 and 231, n.22) refers in the following note to a two-volume edition of *Philosophie der Mythologie* published in Darmstadt in 1966 by the Scholarly Book Society. The source of Schelling's definition hidden in obscurity and twilight in the work of Sanders, Freud, Masschelein and Wirth comes to light and out in the open in Beach's and Vidler's work. As in Freud's essay, Schelling's definition is uncannily present/absent in Vidler's study as it was initially secret, hidden and latent in ellipses and later comes out into the open in a footnote. Schelling's definition of the uncanny is uncanny indeed.

Like Wirth, Teresa Fenichel (2015, pp. 206–207) in her 2015 PhD thesis at Boston College entitled *Uncanny Belonging: Schelling, Freud and the Vertigo of Freedom* cites Beach's translation of Schelling's definition of the uncanny. Unlike Wirth and Vidler, however, she introduces Schelling's definition by saying that he gave it in his 'lectures on mythology and revelation.' This is a grab bag term for Schelling's 'last philosophy' comprising lectures on the philosophy of mythology and revelation, including his 'historical-critical introduction to the philosophy of mythology.'

Beginning in the late 1950s, continuing with his PhD at Columbia University in 1970 and culminating in book publication in 1995, Victor Hayes translated and published a reduction of, and commentary on, Schelling's three series of lectures on the philosophy of mythology and revelation, including the historical-critical introduction. However, Hayes does not include any translations of Schelling's definitions and discussions of the uncanny in the edition of his reduction published in book form in 1995. As a student and teacher of religion and philosophy (and not literature and mythology, unlike Schelling who was a student and teacher of all four, hence his status as a pioneer of transdisciplinary studies) and as an ordained Christian minister, Hayes approaches Schelling's 'last philosophy' largely from a theological point of view and completely hides or obscures the uncanny as an avatar of mythology and polytheism (or paganism, 'i.e., the non-Christian religions' (Hayes, 1995, p. 8)). Schelling's uncanny remains hidden in the unreduced portions of Schelling's lectures on the philosophy of mythology and revelation and does not come to light in Hayes' edition, nor does the context of Schelling's uncanny in his discussion of Homer come out into the open for Hayes.

Beach and Vidler acknowledge that Schelling's definition and mentions of the uncanny invariably occur in the context of his discussion of Homer's epic poems, or more precisely Homeric epic poetry. For Schelling, 'Homer,' or Homeric poetry, is a product of mythology defined by Schelling (2008, p. 9) as Greek 'tales, legends and stories, which in general go beyond historical time,' both into pre-historical time and into what he refers to as *Dictung*, translated as 'poesy.' Without Greek mythology, there would be no Homeric poetry. Homer did not write mythology; mythology wrote Homer. Indeed, Schelling (1989, p. 52) argues in *The Philosophy of Art* that 'mythology and Homer are one and the same.' For Schelling, religion is also a product of mythology. Without mythology, there would be no religion. Religion ignores its mythological roots at its peril and arrogates to itself a role and function as sole access to truth it does not warrant and has no basis for, certainly no scriptural basis.

Scripture is steeped in mythology. The ostensibly monotheistic Judeo-Christian scriptures are saturated with mythology (and so with polytheism, despite the best efforts of the first of the Ten Commandments to keep them at bay), as are the openly polytheistic Buddhist-Taoist scriptures, such as *Ten Thousand Buddhas Sutra* (Fung Loy Kok Institute of Taoism, 2012). Mythology for Schelling (2008, p. 9; his emphases) in the *Historical-Critical Introduction to the Philosophy of Mythology* is 'polytheism' and 'in general the *system of the gods*.' Mythology is neither the system of one god, so nor is religion. Monotheism comes out of polytheism (and not vice versa). Before mythology lies what Schelling (2008, p. 17) calls 'the dark foundry, the first forging place of mythology.' This is the dark, obscure, secret and veiled aspect of the uncanny that comes to light and out in the open in the other aspect of the uncanny. These two aspects of the uncanny come together in mythology. The uncanny is precisely the link and process between these two aspects brought together in mythology.

Mythology and polytheism for Schelling have truth, are a way to truth and even are the truth. Schelling (2008, p. 8) states up front that 'there is indeed truth in mythology.' Schelling (2008, p. 23) restates and refines his initial statement that 'truth is in mythology, but not in mythology as such.' Truth is in polytheism, in difference, not in sameness. Schelling (2008, p. 149; his emphasis) concludes that 'there is truth in mythology *as such*,' because mythology is polytheism, or more precisely what Schelling repeatedly calls 'successive polytheism.' Mythology for Schelling (2008, p. 145) is precisely 'essentially successive polytheism.'

Schelling (2008, pp. 148–149; his emphasis) goes on to explain what he means by 'successive polytheism':

> Polytheism, considered in the entirety of its successive moments, is the way to truth and to this extent truth itself. One could conclude from this: in this way the final mythology, uniting all moments must be the true religion. [. . .] The truth in mythology is initially and especially a *religious* one.

Conversely, the truth in religion is only later and especially a *mythological* one. All religion contains within it an 'untruth.' It is shot through with difference, i.e. potential for polytheism, a tendency that leads to law and prohibition. There is no truth in religion without prior mythological truth – prior historically, ontologically and epistemologically. It is a commonplace of theology that it negates and sublates mythology; by equating mythology with polytheism, Schelling desublates and positivizes theology and monotheism into polytheism.

Monotheism considered in the singularity of one moment and not in the entirety of the successive moments of polytheism prior to and over it is not the way to truth, not the true religion and not primarily the truth. In the contemporary context of the early twentieth century, Schelling's prioritizing of polytheism over and before monotheism may enable the sectors of the monotheisms currently in often violent contention to co-exist. The polytheisms have done so for millennia with mutual respect and cross-fertilization. The 'three religions of China' (Taoism, Buddhism and Confucianism) are a case in point (Fung Loy Kok Institute of Taoism, 2008, 2012).

Schelling (2008, p. 89; his emphases) earlier considered mythology 'as history of the gods, thus the actual mythology, was only able to be produced in life itself; it *had* to be something *lived* and *experienced*.' The truth (of mythology and religion) is something lived and experienced. For Schelling (2008, p. 150) later 'the content of mythology is not an *abstract*-religious one,' but a concrete-religious one. The uncanny is lived and experienced truth as Freud (2003, pp. 144 and 151) shows in his account of getting lost in the red-light district of an Italian town and in his recount of the story of the table with carved crocodiles. The truth is not something believed in in the abstract, a dead, disembodied truth, but a concrete, living, embodied truth.

The way to truth for Schelling lies in nature, or what he calls 'the sensible world.' Nature for Schelling is the first object of knowledge and the site in which the search for God first takes place. 'The immediate object of human knowledge,' Schelling (2008, p. 56) argues, 'remains nature, or the sensible world; God is only the dark, vague goal that is strived for and that is first sought in nature.' Against the view that nature is God, or pantheism, Schelling (2000, p. 31) in *The Ages of the World* counters that 'nature is not God.' Seeking God in nature is polytheistic because mythology is polytheistic, not pantheistic, and truth lies in mythology (and in religion insofar as it is mythological, and not in religion insofar as it neglects or denies its mythological roots and arrogates to itself privileged access to the truth in its scriptures and/or via its priests). Mythology is the way to the truth in life (*contra* Jesus in *John* 14: 6 ('I am the way, the truth and the life'), or Jesus, or the Messiah, is another name, like Homer, for mythology). Mythology for Schelling (2008, p. 155) 'indisputably has the closest affinity with nature.' Theology indisputably does not. Theology has a fraught and problematic non-affinity with nature (see Giblett, 2018). Rethinking religion through mythology and poetics is a way of bringing being and living with the earth into closer affinity with nature (see Giblett, 2011, 2018).

Similarly, art has a fraught and problematic non-affinity with nature. Nature for Schelling (1989, p. 88) is sublime, but 'not sublime in itself' as the sublime is 'actually found in the subject' (p. 90). Adorno (1984, p. 91) bemoaned the fact that Schelling stood at the beginning of 'an almost exclusive concern' in aesthetics with art and that he did not carry on with Kant's 'most perspicuous investigation' into 'the beautiful in nature.' Yet Schelling's view of the sublime is Kantian. Art and nature were mutually exclusive categories for Schelling. Mythology, not art, for him had the closest affinity with nature because mythology arose out of nature whereas art arose out of culture.

Along similar lines to similar ends, one Anglophone Schellingian in Warwick Mules (2014) has taken up Schelling's nature philosophy to argue powerfully and productively for being and living 'with nature.' Mules argues specifically for a demythification of mythological thinking (though it is difficult to see in Schellingian terms how a demythified thinking could have affinity with nature). By contrast, I am arguing (here and elsewhere) for a detheologization and remythification of thinking about, and regaining affinity with, nature, especially via wetlands, swamp serpents and marsh monsters, including crocodiles.

One way to do so leads through Schelling's uncanny and his philosophy of mythology. Matt Ffytche (2013, p. 160) in *The Foundation of the Unconscious: Schelling, Freud and the Birth of the Modern Psyche,* cites Schelling's definition and footnotes '*Philosophie der Mythologie*, 649,' the same page number as Sanders cites. Ffytche later suggests that 'Freud may have looked up the source context of the quotation in Schelling's lectures.' If he had done so, he may have found a kindred spirit interested in literature and mythology. Ffytche's (2013, p. 302) bibliography lists '*Philosophie der Mythologie*, in *Sämmliche Schriften*, Abt. 2, vol. II,' the same reference as Sanders cites (2.2, 649). Sanders' cryptic reference is decoded, the crypt opened and the precise source for Sanders' citation of Schelling's definition comes out into the open. Schelling's definition of the uncanny is uncanny indeed.

Elsewhere, in the *Historical-Critical Introduction to the Philosophy of Mythology*, Schelling mentions the uncanny on several occasions that contribute to an understanding of mythology and polytheism, especially the association between the divine and the monstrous. He links 'religious terror' and 'uncanny feelings' (Schelling, 2008, p. 14), refers to 'the dark and uncanny power of the belief in the gods' (p. 45) and 'the thoughtless dread before something uncanny' (p. 54), and associates the uncanny with the excessive, even monstrous (p. 65). Yet the uncanny does not rate an index entry in this translation and remains relatively hidden in it (p. 239). Although Freud does not make the link that Schelling does between the uncanny, the divine, thoughtless dread, religious terror and the monstrous, he might have done so when he recounted the story in *The Strand* magazine that he read involving a table with carved crocodiles (Freud, 2003, p. 151; Moberly, 1917, 1991). Crocodiles are a vehicle and vector for the uncanny, and for the divine and/or monstrous before they were separated out (as we saw in a previous chapter; see also Giblett, 2009, chapter 2, 2018, chapter 4).

The crocodiles in the story are also associated with the obscure and darkness, qualities of the uncanny for Schelling. In the *Historical-Critical Introduction to the Philosophy of Mythology* Schelling (2008, p. 63) argues that when God created light and darkness, and 'the antithesis' between them as 'the first creation,' 'God calls the light good without naming the darkness evil.' For Schelling (2000, p. 4) in *The Ages of the World*, 'Everything begins in darkness.' Everything was good, including darkness, when God created it, and God loudly declared it to be so in 'repeated assurance,' as Schelling puts it. Some Christian theologians, such as Norman Habel and Claus Westermann, follow suit in their readings of the biblical creation story in *Genesis* 1 (see Giblett, 2018, chapter 1). The darkness associated with the uncanny is not necessarily evil, nor is the monstrous, though the monstrous in Judeo-Christian terms is evil. Cain, the murderous brother of Abel, created evil monsters in the form of Nephilim, or giants, who were 'warriors of old' (*Genesis* 6: 4) and in the form of the marsh monsters

Grendel and his mother in the interpolation of new verses in *Genesis* in the Old English *Beowulf* (see Giblett, 2018, chapter 2). The monstrous is not evil in non-Christian terms. The uncanny is one way of thinking of darkness and the monstrous as not evil.

The monstrous is frightening as it goes back to what was once well known and had long been familiar. It goes back to the first act of fratricide, back to the past. The monstrous crocodiles in the story that Freud condenses is, I have argued elsewhere, a figure for the colonial repressed that returns, and to which the protagonists return, by way of the colonial artefact of the table with carved crocodiles on it (see Giblett, 2009, chapter 2, 2018, chapter 4). The monstrous crocodiles are a vehicle and vector for the uncanny in which the repressed not only returns, but also to which one returns. The table with carved crocodiles on it returns one of the protagonists in the story to the swamp in which a friend was taken by a crocodile. The uncanny not only entails the return *of* the repressed, but also the return *to* the repressed.

Schelling's uncanny is a rich, dark brew which yields nourishment for the body and soul for those who value spirituality and materiality. It provides a progressive understanding of religion in the contemporary world of the contending sectors of monotheisms that might lead to a new appreciation for polytheism. Long may it, and its gods and goddesses, live.

Notes

1 The German text of the extract from Daniel Sanders' *Worterbuch der Deutschen Sprache* cited in Freud's essay is helpfully included as an appendix in *The Standard Edition,* with the '2, 2, 649' cryptic reference to Schelling (Freud, 1955, pp. 252–256).

2 I am grateful to Warwick Mules for this point.

References

Adorno, T. 1984. The Beauty of Nature. In: *Aesthetic Theory*. G. Adorno and R. Tiedemann, eds. C. Lenhardt, trans. London: Routledge and Kegan Paul, pp. 91–115.

Beach, E. 1994. *The Potencies of the God(s): Schelling's Philosophy of Mythology*. Albany, NY: State University of New York.

Benjamin, W. 2015. *On Photography*. E. Leslie, ed. and trans. London: Reaktion Books.

Castle, T. 1995. *The Female Thermometer: Eighteenth-Century Culture and the Invention of the Uncanny*. Oxford: Oxford University Press.

Fenichel, T. 2015. *Uncanny Belonging: Schelling, Freud and the Vertigo of Freedom*. PhD thesis, Boston College.

Ffytche, M. 2013. *The Foundation of the Unconscious: Schelling, Freud and the Birth of the Modern Psyche*. Cambridge: Cambridge University Press.

Freud, S. 1955. The 'Uncanny.' In: *The Standard Edition of the Complete Psychological Works of Sigmund Freud, Volume XVII (1917–1919): An Infantile Neurosis and Other Works*. A. and J. Strachey, trans. London: Hogarth, pp. 217–256. First published 1919.

Freud, S. 2003. *The Uncanny*. D. Mclintock, trans. London: Penguin. First published 1919.

Fung Loy Kok Institute of Taoism. 2008. *A Path of Dual Cultivation: Teachings of the Fung Loy Kok Institute of Taoism*. Toronto: Fung Loy Kok Institute of Taoism.

Fung Loy Kok Institute of Taoism. 2012. *Ten Thousand Buddhas Sutra*. Fung Loy Kok Institute of Taoism, trans. Toronto: Fung Loy Kok Institute of Taoism.
Giblett, R. 1996. *Postmodern Wetlands: Culture, History, Ecology*. Edinburgh: Edinburgh University Press.
Giblett, R. 2008. *The Body of Nature and Culture*. Basingstoke: Palgrave Macmillan.
Giblett, R. 2009. *Landscapes of Culture and Nature*. Basingstoke: Palgrave Macmillan.
Giblett, R. 2011. *People and Places of Nature and Culture*. Bristol: Intellect Books.
Giblett, R. 2016. *Cities and Wetlands: The Return of the Repressed in Nature and Culture*. London: Bloomsbury.
Giblett, R. 2018. *Environmental Humanities and Theologies: Ecoculture, Literature and the Bible*. London: Routledge.
Hayes, V. 1995. *Schelling's Philosophy of Mythology and Revelation: Three of Seven Books Translated and Reduced with a General Introduction*. Armidale: Australian Association for the Study of Religions.
Masschelein, A. 2011. *The Unconcept: The Freudian Uncanny in Late-Twentieth Century Theory*. Albany, NY: State University of New York Press.
McGrath, S. 2012. *The Dark Ground of the Spirit: Schelling and the Unconscious*. East Sussex: Routledge.
Moberly, L. 1917. Inexplicable. *The Strand Magazine*, 54(324), pp. 572–581.
Moberly, L. 1991. Inexplicable. In: *Strange Tales From the Strand*. J. Adrian, ed. Oxford: Oxford University Press, pp. 183–195. First published 1917.
Mules, W. 2014. *With Nature: Nature Philosophy as Poetics Through Schelling, Heidegger, Benjamin and Nancy*. Bristol: Intellect Books.
Royle, N. 2003. *The Uncanny*. New York: Routledge.
Schelling, F. 1989. *The Philosophy of Art*. D. Stott, ed. and trans. Minneapolis: University of Minnesota Press.
Schelling, F. 2000. *The Ages of the World (Fragment) From the Handwritten Remains: Third Version (c. 1815)*. J. Wirth, trans. Albany, NY: State University of New York Press.
Schelling, F. 2008. *Historical-Critical Introduction to the Philosophy of Mythology*. M. Richey and M. Zisselsberger, trans. Albany, NY: State University of New York Press.
Vidler, A. 1992. *The Architectural Uncanny: Essays in the Modern Unhomely*. Cambridge, MA: MIT Press.
Wirth, J. 2003. *The Conspiracy of Life: Meditations on Schelling and His Time*. Albany, NY: State University of New York Press.

5 The uncanny and the work of Walter Benjamin

Like Freud, Benjamin was an exile. Like Freud also, Benjamin was a lapsed or secular Jew. Unlike Freud, Benjamin's concerns were saturated with theology and inflected with historical materialism. Like Freud, Benjamin was concerned with the relationship between the present and the past, and with human's interactions with technology. The uncanny is a concept/metaphor that highlights the differences and similarities between them. The uncanny as the return of the repressed, and the return to the repressed is akin to what Walter Benjamin (1999a, p. 462) called 'a dialectical image' between 'the what-has-been' and 'the fulfilled now.' For him, 'While the relation of the present to the past is a purely temporal, continuous one, the relation to what-has-been to the now is dialectical: is not progression but image, suddenly emergent' (Benjamin, 1999a, p. 462). The 'what-has-been' is not merely (in) the past, but also what has been lost, what is absent or in oblivion, not only past in time, but lost in space, in a place, not only historical, but also geographical (as in the story of the table with carved crocodiles). Benjamin (1999a, p. 464) elaborates further that 'in the dialectical image, what has been within a particular epoch is always, simultaneously, "what has been from time immemorial."'

'The dialectical image' for Benjamin (1999a, p. 474) is 'the primal phenomenon of history.' It is also the primal phenomenon of geography and corporeality. One such dialectical image of the primal phenomenon of history Benjamin develops is drawn from J. J. Bachofen's *Mother Right*, a theological work in some sense as it is about the pagan divinity of the Great Goddess of the swamps where the alligators and crocodiles live. Benjamin (1996, pp. 426–427, 2002, pp. 11–24) discusses Bachofen directly and extensively in an essay and briefly in a review of an Italian book about Bachofen without mentioning in either piece Bachofen's discussion of the swamp world. However, in his essay on the tenth anniversary of Kafka's death, Benjamin (1999b, p. 808) reads Kafka's work in Bachofen's terms of the image of 'the swamp world' and translates it into the language of historical materialism. The swamp world is a primal phenomenon not only of history, but also of geography and corporeality, the grotesque lower earthly and bodily strata (see Bakhtin, 1984; Giblett, 1996, chapter 6, 2008a, chapters 4 and 5). Rather than aesthetic sublimation into earthly transcendence, the swamp world desublimates the earthly and transcendent into slimy and bodily immanence (see Giblett, 1996, especially chapter 2) and into the psychological uncanny and spiritual melancholy (see Giblett, 1996, chapter 7, 2016, pp. 92–96). This is the language of geographical, pre-historical and corporeal materialism of the grotesque lower bodily and earthly strata. It is also the language and laughter of carnival in Bakhtin's (1984) cognate terms. The uncanny, as Dodds (2011, pp. 121–122) points out, is 'something that emerges when carnival, and all that it stands for, becomes repressed,' including, as Dodds goes on to point out, in E. T. A. Hoffmann's gothic horror stories (as we will see in a later chapter).

Benjamin (1999b, p. 809) comments on Bachofen that 'the fact that this stage [of "the swamp world"] is now forgotten does not mean that it does not extend into the present. On the contrary: it is present by virtue of its very oblivion.' The past is not dead; the past is resurrected in the present, the absent brought back to presence, in and by the dialectical image. The 'what-has-been' of the lost swamp world extends into the present by virtue of its very oblivion. The 'what-has-been' of the lost swamp is not only in the past, in history, forgotten in and by the present, but also in the present, in geography, lost in the city now, present by virtue of its very oblivion (which I will show later specifically in relation to Paris).

History and geography have their Angels who preside over the past and the present, over lost and found, present and absent times and places. The Angel of History for Benjamin looks back into the past and the destruction that took place in it with her back turned towards the future and with the winds of progress beating into her face and pushing her into the future. Rather than 'a chain of events that appears before us,' the Angel of History for Benjamin (2003, p. 392) 'sees one single catastrophe which keeps piling wreckage upon wreckage and hurls it at his feet.' The Angel is 'driven irresistibly into the future to which his back is turned, while the pile of debris before him grows toward the sky.' Rather than a series of catastrophes proceeding chronologically one after the other through history, Benjamin's 'Angel of History' sees one single catastrophe of wreckage in time. This single catastrophe, however, occurs not only in time, in history, but also in space, in a place, in geography (as the biblical story shows). The 'Angel of Geography' sees one single catastrophe of wrecked and lost places, such as wetlands dredged, filled and so 'reclaimed,' cities and suburbs set in them and being re-reclaimed by them in storms and floods (and increasingly so in the age of global warming), and cities ruined in destructive cataclysms, such as Sodom and Gomorrah, but more in watery ones than fiery ones.

As well as the Angels of History and Geography – in one case whose divine and eternal point of view of wreckage is unattainable by mere mortals and in the other whose omniscient and omnipresent point of view of wrecked places is also unattainable by mere mortals – 'the Angel of Corporeality' lives bodily present in mind and place, a non-dualistic, holistic way of being probably unachievable by modern human beings. While the Angel of History turns 'the threatening future' into 'a fulfilled now' as Benjamin (1996, p. 483) puts it, the Angel of Geography turns threatening places into a fulfilled here and the Angel of Corporeality turns the threatening mind and/or body into a fulfilled, whole mind-body. From this point and place in space-time, these Angels look over disastrous places and ruined bodies and minds, as well as look back at the disastrous past with mixed fear of, and nostalgia for, the past and lost places. They are not afraid and they are in no danger of being turned into a pillar of salt, like Lot's wife, for they are not fixated on the past. They face the past, present, places and the future with fear and hope in embodied minds and mindful bodies.

Unlike Freud, Benjamin retained many concepts of theology as Benjamin (1999a, p. 471) himself proclaimed in his famous comment that 'my thinking is related to theology as blotting pad is related to ink. It is saturated with it. Were one to go by the blotter, however; nothing of what is written would remain.' For Scholem (2012, p. 187), Benjamin's 'insights are those of a theologian marooned in the realm of the profane' and who 'translates them into the language of historical materialism.' Yet Benjamin marooned like the survivor of a shipwreck in the realm of the profane and secular on the desert island of high modernity in the late nineteenth and early twentieth centuries still found a home in the realm of the sacred (and not just in theology) and in the language of historical materialism, perhaps unlikely bedfellows, but ones which he brought together successfully in a new language of materiality and spirituality.

Scholem (2012, p. 197) goes on later to note the presence in the general of 'the messianic idea' among 'the Jewish categories' that Benjamin introduced into his writings and 'upheld to the last.' Scholem (2012, pp. 101–126) does not specifically mention here (though he does in the following essay in the same collection devoted to 'Walter Benjamin and his Angel') the Jewish theological category of the messianic now, or *Jetztzeit.* This is a concept that Benjamin (2003, p. 395) developed coming out of traditional Judeo-Christian theology (with its concept of *Kairos* derived from Greek thought and probably via the Christian apostle and New Testament epistolary writer, St Paul) and combined with Marxist eschatology to indicate a messianic irruption in the present that blasts open the continuum of history (in a word, revolution).

In Benjamin's (1996, p. 483) own terms, in order 'to turn the threatening future into a fulfilled "now," the only desirable telepathic miracle, is a work of bodily presence of mind.' This miracle is performed by the Angel of Corporeality. For Benjamin the body is, or can be, in the fulfilled present in the mind and in the now going into the future. This was certainly the case for Benjamin (1996, p. 483) as he goes on to suggest that 'primitive epochs, when such demeanor was part of man's daily husbandry, provided him, in the naked body, with the most reliable instrument of divination. [. . .] What would have become a portent of disaster he binds bodily to the moment, making himself the factotum of his body.' 'Primitive man' does all kinds of work *for* his body (dance, ritual, tai chi, yoga, etc.), whereas 'modern man' does all kinds of work *with*, or *on*, his or her body (body-building, diet, exercise, sport, etc.). The traditional human being works his or her whole body and not just parts of it. Rather than the body being a portent of disaster in the future (of disease, decay and death), the body sutures him/her to the present and fulfils the now and the here.

Benjamin later commented on 'modern man' and his relationship to his own body. Writing on the tenth anniversary of Kafka's death, Benjamin (1999b, p. 806) argues that in Kafka's *The Castle*, 'just as K lives in the village on Castle Hill, modern man lives in his body; the body slips away from him, is hostile towards him. [. . .] Exile – his exile – has gained control over him [. . .] the most forgotten alien land is one's own body.' Benjamin and Stefan Zweig lived this exile from their own bodies and a double exile from their own homeland (see Giblett, 2018, Chapter 8). By contrast, for traditional human beings the most fondly remembered homeland is one's own body. Exile is not just a matter of leaving one's home and homeland, but is also experienced in relation to one's own body, something that Benjamin wrestled with his whole life. K is not at home in his own body; the castle is not a home. Being at home includes being at home in one's own body and place. For traditional humans, one's land is one's own body and vice versa (see Giblett, 2008a). They are devotees, recipients and hosts of the Angels of Corporeality, Geography and History.

Benjamin was also concerned with modern technologies of mechanical reproduction, principally photography and cinema, and with the changing perceptions of time and space, as well as the training of consciousness and the senses, associated with them. Photography can be considered as literally and simply 'light-writing.' This is good etymology but problematic aetiology for it implies the question of what sort of writing? For Virilio (1989, p. 81), 'Photography, according to its inventor Nicéphore Niépce, was simply a method of engraving with light, where bodies inscribed their traces by virtue of their own luminosity.' Yet in inscribing its traces the engraved body was engaging in a double process split between the writing of inscription and the writing of the trace (see Giblett, 1996, chapter 3, 2008b, chapter 1). The body engraved in and by the photograph always leaves its traces, just as living for Benjamin is 'a leaving of traces' (Adorno and Benjamin, 1999, p. 104; see also Benjamin, 1973a, p. 169). Photography for Benjamin (1973a, p. 48, 2003, p. 27) 'made it possible for the first time to preserve permanent and unmistakable traces of a human being,' but s/he is not necessarily inscribed in the process. Photography is light-writing, but the writing of photography is split between writing as inscription and writing as trace. Benjamin (1999b, p. 512, see also 1979, p. 244) argues that 'the first people to be reproduced entered the visual space of photography with their innocence intact – or rather, without inscription.' In other words, they entered it, and their image was reproduced, or traced, with their aura.

Photography as trace and without inscription is auratic. Benjamin found in early photographs the qualities of aura and the uncanny. Benjamin (1999b, p. 518, see also 1973b, p. 222 and 224, 2002, pp. 103–105, especially 123 n.5, 2003, pp. 253–256) defines aura as 'a strange weave of space and time: the unique appearance or semblance of distance, no matter how close it may be.' To what does 'it' refer? Grammatically it refers to 'distance,' but how can distance be close, or faraway, for that matter? *That* is aura, the closeness or proximity of distance when distance is overcome, but still maintained. Aura is the experience of an object looming up and receding at the same time. In collapsing time and space together, early photography produced an auratic 'presence in time and space' (Benjamin, 1973b, p. 222).

Aura is akin in this respect to Schelling's and Freud's concept of the uncanny, a strange, spectral presence/absence. Just as the uncanny is the return *to* the repressed content or portion of the unconscious (see Giblett, 1996, chapter 2), so the auratic is the return to the repressed of what Benjamin (1979, p. 243, 1999b, p. 512, 2002, p. 117, 2003, p. 266) called 'the optical unconscious.' And just as psychoanalysis brought the former to light, photography for Benjamin brings the latter to light – literally:

> For it is another nature that speaks to the camera than to the eye: other in the sense that a space informed by human consciousness gives way to a space informed by the unconscious. . . . It is through photography that we first discover the existence of this optical unconscious, just as we discover the instinctual unconscious through psychoanalysis.
>
> (1979, p. 243, 1999b, pp. 510–511)

Just as the uncanny is by definition invisible, but is made sensible, especially through the sense of smell (as we saw in chapter 2), so the auratic is the invisible made visible.

Aura is also similar to Freud's concept of symptom in which the surface of the body of the patient (including their behaviour and clothing) bears and manifests the traces of their psychopathology (see Giblett, 1996, chapter 4). Aura operates in the circuit of the uncanny and symptomatic: simultaneously returning to the repressed to return the repressed to the surface of the body. Aura is the expression in old photographs of a profound and unique moment in time and a place in space; aura is more than a mere event, but the imbuing of an event with ritual significance that transcends time and space. It is the trace of the performance of a body in eternity. Commenting on a *c.*1850 photograph of Schelling, Benjamin (1999b, p. 514) argues that 'the creases in people's clothes have an air of permanence.' Even the evanescent creases in their faces have a similar air of permanence. Aura is the play of permanence and evanescence. Benjamin (1973b, p. 228, 2002, p. 108, 2003, p. 258) suggests that 'for the last time the aura emanates from the early photographs in the fleeting expression of a human face.' Photography kills the object in photographing it, but the photograph always bears the traces of the living body of the subject, whether it be human, animal, vegetable or mineral, though it will not necessarily convey its aura.[1]

Benjamin (1999b, pp. 327–328) distinguishes three aspects of genuine aura: 'First, genuine aura appears in all things, not just in certain kinds of things.' Aura is the sacral quality with which all objects (including everyday objects and subjects) are imbued in traditional and pre-modern cultures. A vestige of that quality lives on in modern cultures in the fetishism and phony spell of commodities. 'Second, the aura undergoes changes.' Aura is not fixed or eternal; it can wax and wane. The auratic significance with which objects are imbued ebbs and flows without ever becoming totally bereft of it. 'Third, the characteristic feature of genuine aura is ornament, an ornamental halo, in which the object or being is enclosed as in a case.' The auratic object is framed or contained. It is marked off decoratively and spatially from other objects with which it shares sacrality on a continuum.

These three aspects of aura are also found in the uncanny. Objects, whether they be real crocodiles, carved crocodiles or virtual crocodiles, are vehicles and vectors for the uncanny. They are imbued with a sacral quality. The uncanny lives and dies, blooms and withers, waxes and wanes, like real crocodiles living in swamps, crocodiles carved in wood from dead trees or virtual crocodiles resurrected in artefacts and art works. The uncanny is also veiled, just as the aura object is haloed and ornamented. The table with carved crocodiles is also marked off decoratively and spatially from other objects.

The fundamental distinction between the auratic and the inscriptive does not correlate to that between nature and culture, nor between orality and literacy, but between the cultures of first, or worked, nature and of second, or worked over, nature (see Giblett, 2011, chapter 1). Aura is pervasive in the former, vestigial in the latter. For Benjamin (1999a, p. 362), 'The decline of the aura and the waning of the dream of a better nature . . . are one and the same.' Both go hand in hand with the commodification of nature, the photograph and photography. Nature is drained of sacral significance and imbued with capitalist value, just as the aura of human subjects and other objects declines – and just as the photography of events rises. For Heidegger, the fundamental event of the modern age is the conquest of the world as picture. Photography contributes not only to that conquest but also to the fundamental event of the hypermodern age: the conquest of the world as commodity.

Aura declines as images of objects are reproduced and as exposure times decrease 'from Niépce's 30 minutes in 1829 to roughly 20 seconds with Nadar [in] 1860' (Virilio, 1994, p. 21). Aura in photographs is a function partly of the long exposure times and slow lenses and shutter speeds of old cameras, and partly of the subject's intact innocence in old photographs. Yet the two go hand in hand as the snapshot produced by cameras with short exposure times and fast lenses and shutter speeds produces both a guilty subject who knows he/she is having his/her photograph taken and a commodified object of the photograph that is bought and can be resold. Capitalist commodities can be decorative, but mass reproduction does not (and cannot) mark them off spatially from other mass-produced objects. Benjamin (1999a, pp. 337 and 343) argues that 'for the decline of the aura, one thing within the realm of mass production is of overriding importance: the massive reproduction of the image.'

The aura of a work of art, in Benjamin's (1973b, p. 223, 2002, p. 104, 2003, p. 254) words, 'withers in the age of mechanical reproduction.' Aura is living matter so it can wither, or 'shrivel,' as Benjamin (1973b, p. 233) elsewhere puts it, like a plant. The photograph is commodified, sold and bought, like the book or painting, and so is dead matter. The photograph is a commodity, or dead matter, by virtue of light-writing the outlines of object on a flat surface. Living matter sublimated into air is transformed into dead matter (see Giblett, 1996, Figure 1).

The new photograph is split between its status and function as inscription and as commodity. Photography is split between the old photograph and the aura it possesses, and the new photograph with its vestiges of aura and traces of the object.

Aura is a kind of divine fire, the living fiery breath of God, or the Holy Spirit, that 'He' breathed into everything in creation according to the biblical book of *Genesis*. Photography is an attempt not only to wrest fire from the gods, but also to reproduce it. Neale (1985, p. 25) points out that 'there existed a whole theology of light as the trace of the noumenal in the real, as the mark of spirit in matter.' Divine light was incarnated in the photographic body of the son/sun of God. Solid matter was sublimated into the sublime image in the *camera obscura* only to be incarnated and inscribed in the sublimate of the surface of the photograph, and later animated in cinematic film. Photography was the logical next step to the *camera obscura* whose image was, as Neale (1985, p. 20) puts it, 'fleeting, intangible, evanescent.' Photography in its beginnings was concerned primarily with recording the event of the shadow rather than the outlines of the object. It was more concerned with the index of the presence of the object than with the object itself.

Cinema was only to heighten and broaden these aspects of photography. Cinema is mental machinery. The machine is mental to the point that Benjamin (1973a, p. 132, 1973b, p. 177, 2003, p. 328) argues:

> Technology has subjected the human sensorium to a complex kind of training. . . . In a film, perception in the form of shocks was established as a formal principle. That which determines the rhythm of production on a conveyor belt is the basis of the rhythm of reception in the film.

The early 'cinema of attraction' (Gunning, 1986, 1989) was a cinema of shock ('rapid cutting, multiple camera angles, instantaneous shifts in time and space') (Wolin, 1982, p. 233). The shock of the new came at the price of the loss of the old, the traditional, the auratic. For Benjamin (1973a, p. 154, 1973b, p. 196, 2003, p. 343), 'The price for which the sensation of the modern age may be had [is] the disintegration of aura in the experience of shock.' The instantaneity of the shock administered by the moving pictures of cinema (and later television) disintegrated the protraction of the aura. Just as shorter exposure times and brief posing for the snapshot brought about the demise of the lengthy sitting for the auratic early photos, so the production of shocks on the assembly line of the cinema diminished the auratic presence of objects. Just as the trace and aura for Benjamin are distinct in photography, so are shock and aura in cinema.

The cinema of shocks arose out of advertising and the new pictorial representationalism instigated by print and lithography. For Benjamin (1979, p. 89, 1996, p. 476):

> Today the most real, mercantile gaze into the heart of things is the advertisement. It abolishes the space where, and tears down the stage upon which, contemplation moved and all but hits us between the eyes with things as a

> car, growing to gigantic proportions, careens at us out of a film screen. And just as the film does not present furniture and facades in completed forms for critical inspection, their insistent, jerky nearness alone being sensational, the genuine advertisement hurtles things at us with the tempo of a good film.

The early 'cinema of attractions,' such as *Arrival of a Train at a Station* that depicted just that, produced not only shock, but also such terror that audiences reportedly ran screaming from the cinema. Such a response was not merely indicative of a naïve realism in the audience that mistook the pictorial representation of the thing for the thing itself, that collapsed realism and reality, but also of a terror at a thing grown to such monumental or monstrous proportions that it towered above and threatened to engulf the minuscule members of the audience, the terrifying disproportion between the sublime machine of the train in the cinema and what Benjamin (1973b, p. 84, 2002, p. 144, 1999b, pp. 731–732) calls 'the tiny, fragile human body.'

The power of 'the movies' lay not just in moving pictures, in the ability of the cine-camera to take pictures continuously (unlike the photographic camera), but also in the moving camera, in the ability of the camera to move and take pictures whilst it is doing so. Benjamin (1999b, p. 17, see also 1973b, p. 238 and 2002, p. 117; his emphases) commented:

> With film, a *new realm of consciousness* comes into being. To put it in a nutshell, film is the prism in which the spaces of the immediate environment – the spaces in which people live, pursue their avocations, and enjoy their leisure – are laid open before their eyes in a comprehensible, meaningful, and passionate way. In themselves these offices, furnished rooms, saloons, big-city streets, stations, and factories are ugly, incomprehensible, and hopelessly sad. Or rather, they were and seemed to be, until the advent of film. The cinema then exploded this entire prison-world with the dynamite of its fractions of a second, so that now we can take extended journeys of adventure between their widely scattered ruins.

As the adventure-journey through ruinous landscapes is the hallmark of Romanticism, cinema created a romanticized, aestheticized and touristic landscape out of the ordinary and everyday. The cinema-goer is a tourist of the everyday; cinema is tourism of the everyday and tourism for the everyday. The cinematograph, Virilio (2000, p. 23) argues:

> was a substitute for human vision which not only flouted time (thanks to the illusion known as persistence of vision), but also flouted the distances and dimensions of real space. The cinema was, in fact, a new energy, capable of carrying your gaze to other places, even if you yourself did not move.

The gaze was carried to other places whilst the body was still sitting in one place. The gaze was mobilized and the body immobilized, even demobilized.

Whereas in landscape painting the Claude glass compressed the country landscape within a concave mirror to make it into a static, painterly object, cinema exploded the city landscape with retinal retention into a continuous stream of fragmentary images. Whereas the Claude glass implied and addressed a single viewer viewing from a single point of view, cinema addressed and positioned an audience seeing the same thing from the same point of view. For Benjamin (quoted by Virilio, 1994, p. 21), 'Cinema provides *matter* for simultaneous collective reception.' This statement implies the question of, which matter? What sort of matter? Virilio (1994, p. 21) answers this question by stating that 'the *matter* provided and received in collective, simultaneous fashion by cinemagoers is light, the speed of light,' not the matter of the solid object filmed for that has been sublim(at)ed into light, become a virtual object and then deposited in the sublimate, the solid object, of film.

Including the body of the star. The heavenly bodies of cinema stars were described by Pirandello (quoted by Virilio, 1989, p. 15; see also Benjamin, 1973b, p. 231, 2002, p. 112, 2003, p. 260) as being, 'so to speak subtracted, suppressed, deprived of their reality, of breath, of voice, of the sound they make in moving about, to become only a dumb image, which quivers for a moment on the screen and disappears, in silence.' The body of the star is reduced to mute visual image that burns with a fleeting, incandescent light, and then dies. Their unique presence in time and space, their address to the ear, the aural, their aura, has gone, especially and quite literally in the early silent films.

Benjamin (1973b, p. 231, 2002, p. 112, 2003, p. 260) goes on to elaborate on the situation outlined by Pirandello: 'For the first time – and this is the effect of the film – man [*sic*] has to operate with his whole living person, yet forgoing its aura. For aura is tied to his presence; there can be no replica of it.' The star system is for Benjamin the precise counterpart and exact compensation for this loss of aura in film. Benjamin (1973b, p. 233, cf. 2002, p. 113, 2003, p. 261) argues:

> The film responds to the shrivelling of the aura with an artificial build-up of the 'personality' outside the studio. The cult of the movie star, fostered by the money of the film industry, preserves not the unique aura of the person, but the 'spell of the personality,' the phoney spell of a commodity.

Aura is the sacral quality of an object present in time and space; the star is the fetishized quality of a commodity cancelling time and space. The star is sublimated into a transcendental realm between solid matter and heady ideality; aura is desublimated into an immanent realm between solid matter and slimy traces.

The star made art into commodity and paved the way for art to be made into politics. As Johnson (1988, p. 6) eloquently puts it, Benjamin 'pointed to the creation of movie stars and the building up of the cult of the personality by film studios as an attempt to replace the desiccated aura of the art object with the artificial aura of the commodity.' Film studios sowed the wind of the movie star only to reap the whirlwind of the cult of personality in the totalitarian dictator, Fascist and 'Communist' (State Capitalist) alike. For Benjamin (1973b, pp. 243–244, 2002, pp. 121–122, 2003, pp. 269–270), 'The logical result of Fascism is the introduction of aesthetics into political life' to which 'Communism responds by politicising art.' The artificial aura of the commodity superseded the dried-up aura of art, and religion, and substituted commodity fetishism for sacrality, dead matter for living being. Capitalism aestheticizes commodities. The modern conquest of the world as picture (as Heidegger put it) and the hypermodern conquest of the world as commodity ultimately come together, and come together in modern, industrial warfare culminating so far for Benjamin writing in the 1930s with World War I.

Benjamin (1999b, p. 732) argued that in what he called 'positional warfare' of WWI:

> a generation that had gone to school in horse-drawn streetcars now stood in the open air, amid a landscape in which nothing was the same except the clouds and, at its centre, in a force field of destructive torrents and explosions, the tiny, fragile human body.

Benjamin took his own life in 1940 before he could remark about the next generation that had gone to school in electric streetcars (trams) and steam railways who in the strategic warfare of the Second World War now sat huddled in the closed air of the bomb shelters amid a landscape of ruins in which they could not even see the clouds, but still at its centre in a force field of greater destructive torrents of fire and exploding bombs, was the still tiny, fragile human body. Yet whereas the tiny, fragile human body that Benjamin remarked upon in the First World War trench warfare in the country was the soldier body, the tiny, fragile human body in the Second World War strategic bombing of the city was the civilian body.

Benjamin (1979, pp. 103–104, 1996, p. 486) noted in the 1920s how, under the conditions of technologically advanced modernity beginning around the time of the First World War:

> human multitudes, gases, electrical forces were hurled into the open country, high frequency currents coursed through the landscape, new constellations rose in the sky, aerial space and ocean depths thundered with propellers, and everywhere sacrificial shafts were dug in Mother Earth. This immense wooing of the cosmos was enacted for the first time on a planetary scale, that is, in the spirit of technology.

Benjamin (1979, p. 104, 1996, p. 487) continues by bemoaning the fact that:

> the mastery of nature, so the imperialists teach us, is the purpose of all technology. But who would trust a cane wielder who proclaimed the mastery of children by adults to be the purpose of education? Is not education above all the indispensable ordering of the relationship between generations and

> therefore mastery, if we are to use this term, of that relationship and not of children? And likewise technology is not the mastery of nature but of the relation between nature and man [*sic*].

The traditional city is not the master of nature but of the relation between nature and humans, whereas the modern city aspires to the mastery of nature, but ends up being mastered, or monstered, by nature in the age of global warming/climate change with increased and higher flooding (see Giblett, 2018, chapter 9).

Naples is the traditional city for Benjamin that exemplifies the mastery of the relationship between nature and humans (and not the mastery of nature). Benjamin's essay or travel diary about Naples, what Eiland and Jennings (2014, pp. 210 and 211) call 'the first of Benjamin's memorable city portraits,' is for them 'uncannily alive to both the city's wretchedness and its glory.' In his portrait of Naples, Benjamin noted that 'the city is craggy' and has 'grown into the rock' on which it is built and from which it is built. Benjamin presses the porous quality of this rock into service as a metaphor for the architecture and spaces of the city. Porosity becomes 'the defining feature of the city' (Eiland and Jennings, 2014, p. 211) literally and materially as well as metaphorically and culturally. Commenting in a radio talk on everyday life and festival days in Naples, Benjamin (2014, p. 151) notes that 'what is remarkable is how the two blend into each other.' Porosity is also literally and materially the defining feature of the city and its architecture partly because of the porous stone out of which it is built. In his essay about Naples, Benjamin (1979, p. 169) notes that 'as porous as this stone is the architecture. Building and action interpenetrate in the courtyards, arcades, and stairways. In everything they preserve the scope to become a theatre of new, unforeseen constellations.' Arcades and courtyards are important city sites for Benjamin (in Paris and Berlin), and constellations are an important astronomical figure for him for bringing seemingly disparate aspects together. Arcades and courtyards are liminal, porous places where inside and outside, the private and the public, fecundly and procreatively interpenetrate and invaginate to give birth to new space-time constellations, just as wetlands are liminal, porous places where solid and liquid, earth and water also fecundly and procreatively interpenetrate and invaginate to give birth to new space-time constellations.

Benjamin (1979, p. 170) goes on to suggest on the following page that 'porosity results not only from the indolence of the Southern artisan, but also, above all, from the passion for improvisation, which demands that space and opportunity be at any price preserved. Buildings are a popular stage.' Wetlands are also a popular stage on which the performance of divine and biological improvisation takes place and on which some cities are built, thereby usually destroying them in the inscription of the city in grids of streets and blocks, and the erection of monuments, on their blank surface. Their porosity is improvisation par excellence resulting from the work of the divine artisan (indolent, passionate or otherwise) who mixed, and biology that mixes, earth and water and thereby created space and time, and opportunity for the performance of improvisation.

The city is uncanny, especially Paris and Berlin. In a radio talk broadcast in 1930, Benjamin (2014, p. 50) referred to 'the uncanny Berlin of a century ago' Paris, like many other cities (including Berlin), was built beside a wetland, the uncanny place *par excellence*. Paris had a covert, dramatic relationship with the city's swampy history in the nineteenth century. The repressed swampy beginnings of Paris, and any other city for that matter built on or by a wetland, return repeatedly in its history, particularly in tropes. For example, d'Aurevilly (cited by Benjamin, 1973a, p. 25) invokes 'the sediment of rancour which has accumulated' in cities. Paris and other cities were founded on the sediment of its wetlands, and this repressed foundation returns as a rich source of metaphor for its cultural life, usually, as here, in the pejorative register. Rather than the Seine being a green, greasy river, Auguste Barbier (cited by Benjamin, 1999a, p. 739) figured Paris as 'an infernal vat/ . . . Ringed by three bends of a muddy yellow river.' Paris as hell at whose centre resides the devil was a commonplace of nineteenth-century figuration of this and other, especially swamp, cities.

The buried wetland history of Paris as Lutetia also returns in Benjamin's figuring of the Paris of Baudelaire's poems as 'a sunken city, and more submarine than subterranean' (Benjamin, 1973a, p. 171). Benjamin goes on to refer to 'the chthonic elements of the city,' such as 'its topographical formation, the old abandoned bed of the Seine.' Paris rests, or floats, on the old bed of the Seine. Lutetia is remembered in the name of the Hotel Lutetia that opened in Paris in 1910 and is still operating. Benjamin (1999a, p. 516) acknowledges 'the unconquerable power in the names of streets, squares, and theaters, a power which persists in the face of all topographic replacement.' How much more so does the unconquerable power of the name of Lutetia not only persist in the name of a hotel, but also endures in the massive topographic replacement of the filthy marsh of Lutetia by the city of Paris?

Although Benjamin (1999a, p. 83) was aware of the Roman name for the city as *Lutetia Parisorum*, he does not discuss the meaning of the word Lutetia (his editors do that), nor note that the city was founded in and on the swamps of Lutetia, even though Haussmann (the blaster of boulevards) and the novelist Balzac do, both of whom Benjamin cites repeatedly in *The Arcades Project*, though necessarily selectively. Haussmann's project is even figured in terms of engineering the urban landscape figured as wetlandscape, such as when Jean Cassou (cited by Benjamin, 1999a, p. 793) describes how 'Haussmann built broad, perfectly straight avenues to break up the swarming, tortuous neighbourhoods, the breeding grounds for mystery . . ., the secret gardens of popular conspiracy.'

Just as the swamp was considered to be the breeding ground for malaria in the mistaken view of the miasmatic theory of disease, so the slum was considered the breeding ground for class revolt and a pretext for the boulevarding of the city. Along similar lines, Aragon (1994, p. 14) regarded Baron Hausmann as the importer of 'the great American passion for city planning' whose 'redrawing the map of our capital in straight lines' would 'soon spell the doom of these human aquariums' of the arcades. The glass enclosures of the arcades were for

Aragon, as they were for Benjamin (1999a, p. 87), 'the true sanctuaries of a cult of the ephemeral, the ghostly landscape of damnable pleasures and professions.'

These qualities of the arcades were associated for Benjamin with phantasmagoria. Writing on Benjamin and phantasmagoria, Cohen (1993, p. 219) describes how, 'in nineteenth century usage[,] this term designated both a form of magic lantern show (illustrated in Cohen, 1993, p. 216) and a psychological experience when the distinction between subject and object breaks down.' As a result of this breakdown, the subject becomes object in the reified relationship between the worker and work, consumer and commodity. Cohen (1993, p. 222) goes on to argue that in 'the magic . . . of commodity fetishism . . . social relations . . . take on the phantasmagorical form of relations between things.' For Benjamin (1999a, p. 14) 'reifying representations . . . enter the universe of a phantasmagoria' in which the consumer is placed under what Benjamin (1973b, p. 233) called 'the phony spell of a commodity.'

The phantasmagoria for Benjamin has its representative human figure in the *flâneur* and manifest expressions in his appearance and behaviour. For him, 'The *flâneur* abandons himself to the phantasmagoria of the marketplace' and Haussman is 'the champion' of 'the phantasmagoria of civilization' whose 'manifest expression' is 'his transformation of Paris' (Benjamin, 1999a, pp. 14–15). An earlier expression and champion is Tsar Peter's transformation of Petersburg (see Giblett, 2016, chapter 7). In the case of both cities, an abject wetland is transformed into a phantasmagorical city in which the subject is objectified by commodity capitalism in modernity. For Benjamin (1999a, p. 26), 'modernity' is 'the world dominated by its phantasmagorias' that he saw exemplified in the arcades and the *flâneur*. This is no less so than in the founding and building of the cities of modernity, such as Petersburg, and in the capital of modernity in Paris, built in pre-modern and abject swamps and marshes. Here the distinction between subject and object has not yet been constituted; here the phantasmagoria of commodity capitalism has not yet broken down the distinction between them in the reification of subject into object under the phony spell of a commodity; here the magic of the marsh and swamps casts its binding spell before the city drains and fills them and creates its phantasmagorias, not least of itself.

Benjamin figured Paris as a labyrinthine whore. In *The Arcades Project* Benjamin (1999a, p. 523) cites a nineteenth-century account of 'the true Paris [which] is by nature a dark, miry, malodourous city . . . swarming with blind alleys . . . and . . . with labyrinths that lead you to the devil.' The city for Benjamin is ultimately a labyrinthine female body. Indeed, for Benjamin (1999a, p. 519) the darkness of the streets 'greatly resembles the lap of a whore.' The sexually and spatially repressed is the bodies of the prostitutes embodying what Solnit (2000, p. 209) calls his 'transformation of city into female body.' Benjamin (1999b, pp. 141–143) also figures 'Paris as Goddess.' In writing about Paris, especially in his labyrinthine, monumental and monstrous *Arcades Project*, a *femme fatale* of a book, Benjamin was primarily interested in Paris as 'the capital of the nineteenth century' as he put it, and in critiquing, as Baudelaire put it, 'the goddess of Industry' (cited by Benjamin, 1973a, p. 79) figured as an orally sadistic monster with her 'jaws' that consume rather than celebrating the 'goddess of the city,' as Benjamin also put it (1999b, p. 143), in the preceding centuries going back to its beginnings as Lutetia as an orally satisfying mother of the marsh.

Traces of the swamp nevertheless can be found in the *Arcades Project*. Baudelaire's writings are one of the central proof texts for Benjamin with good reason. Baudelaire for Benjamin (1973a, p. 81) performs the labours of Hercules to 'give shape to modernity.' Benjamin does not mention that one of Hercules' labours was to kill the brazen-beaked Stymphalian birds that lived in a swamp and are figured as orally sadistic monsters. Benjamin's Herculean labour was to kill the monstrous goddess of industry, rather than to celebrate the Lutetian mother goddess of the marsh. The former gave birth to *The Arcades Project* somewhat surprisingly described by Pike (2007, p. 67) as 'the most sustained meditation to date on the devil (inter alia) in the nineteenth-century metropolis.' The monstrous goddess of industry for Benjamin is certainly a satanic figure of ambiguous and dubious gender; Satan for Milton was certainly a swamp monster of slimy origins (Giblett, 1996, pp. 183–185). The slimy and swampy origins of Paris (as with other cities) make it an ideal location for the engendering of satanic and monstrous industry that consumes the earth, its inhabitants and resources. France for Papini (cited by Pike, 2007, p. 67) is 'the promised land of Satan' and Paris is its capital.

Benjamin's account of Paris focuses on the culture and history of the city in the nineteenth century, a big enough undertaking as the sheer bulk of *The Arcades Project* demonstrates. Yet his lack of attention to the matrifocal prehistory of the site for the city, especially when it re-emerges in the literature of the nineteenth century, is somewhat surprising given his interest in Bachofen's work on 'mother right,' or the great goddess, expressed both in the *Arcades Project* (Benjamin, 1999a, p. 361) and elsewhere in his review of Bachofen's book on the topic (Benjamin, 1996, pp. 426–427) and in his essay on Kafka written on the tenth anniversary of his death (Benjamin, 1999b, pp. 808–809).

If Benjamin had been aware of the beginnings of Paris in the swamps of Lutetia, or the meaning of the name, he may have made the connection between the work of Bachofen, the history of Paris as manifested in *The Arcades Project* (Benjamin, 1999a), and its pre-history. Regrettably he does not make this connection. Bachofen, as Benjamin points out, discusses the 'hetaeric stage' of the pre-patriarchal or matrifocal 'Mother Right,' or 'Great Goddess,' of 'the swamp world,' when the world was swamp (as Benjamin puts it in writing about Kafka; Benjamin, 1999b, pp. 808–809). Benjamin (1999b, p. 809) comments that 'the fact that this stage is now forgotten does not mean that it does not extend into the present. On the contrary: it is present by virtue of its very oblivion.'

Nineteenth-century French literature manifests not only this forgetting as the Great Goddess is not mentioned in it, but also this extension into the present as Lutetia is frequently invoked. Benjamin does not make the connection between the two. The past of what Benjamin calls 'the dark, deep womb' of what Bachofen calls (citing Arnobius) 'dirty voluptuousness (*luteae voluptates*)' (Benjamin, 1999b, p. 809) is present in the past name of Paris, *Lutetia*, not merely as dirty voluptuousness, but as dirty *swampy* voluptuousness, or, as Gilles

Deleuze (1989, p. 52) puts it in his book on Bachofen, 'the lustful chaos of primeval swamps.' The reminder in the present of the past name of Paris as Lutetia 'takes us back,' as Benjamin puts it, to the past of the swamp world in which Paris began, the world of the Great Goddess, though Benjamin forgets about this too. The present figuring of Paris as Goddess takes us back to when Lutetia the swamp was the Goddess, and not just merely figured as such.

The repressed, rich dark past of Paris swamps also returns in a one-act vaudeville performance in which, as Benjamin (1999a, p. 56) puts it, 'Lutèce emerges from the bowels of the earth, at first in the guise of an old woman.' The editors of *The Arcades Project* briefly note that 'Lutèce' is the 'Roman name for Paris' (Benjamin, 1999a, p. 959, n.12), whereas it was previously the Celtic name for the wetland place. Similarly, in their 'Guide to Names and Terms' the editors of *The Arcades Project* gloss 'Lutèce' as 'ancient name for Paris. From Latin *Lutetia* ("city of mud")' (Benjamin, 1999a, p. 1038). However, as Benjamin (1999a, p. 83) says later (and as we have seen), 'Lutetia Parisorum,' not just Lutetia, is 'the old Roman city.' In the vaudeville performance that Benjamin paraphrases Lutetia is an old woman; in Bachofen's terms, she is the Great Goddess who lives in the bowels of the earth, the swamps.

One of the ways in which for Benjamin the hetaeric stage extends into the present is as a point of ecological critique of the present. For him, 'The murderous idea of the exploitation of nature, which has ruled over things since the nineteenth century . . . could have no place so long as the prevailing image of nature was that of the ministering mother, as reflected in Bachofen's conception of matriarchal societies' (Benjamin, 1999a, p. 361). The green or eco-critical Benjamin has not been noted or commented upon much by Benjamin scholars, nor by eco-critics and environmental theorists either for that matter (see, for exceptions, Ebbatson, 2013; Giblett, 2009, 2011; Mules, 2014).

How ironic that the rule of 'the murderous idea of the exploitation of nature' should have its beginnings for Benjamin not only in the nineteenth century, but also in Paris as 'the capital of the nineteenth century' whose original name of Lutetia is precisely that of 'the ministering mother' of the swamp world. Lutetia in Benjamin's terms is 'a dialectical image' for, as he goes on to elaborate, 'while the relation of the present to the past is a purely temporal, continuous one, the relation to what-has-been to the now is dialectical: is not progression but image, suddenly emergent' (Benjamin, 1999a, p. 462). Lutetia is the dialectical image of the what-has-been of Paris as marsh and of the now of recognizability of Paris as metropolis, between the filthy marsh and the miry city. Marsh and metropolis come together dialectically in the image of Lutetia, of Paris as marsh metropolis.

Benjamin (1999a, p. 464) elaborates further that 'in the dialectical image, what has been within a particular epoch is always, simultaneously, "what has been from time immemorial."' Lutetia has been from time immemorial and has been in the particular epoch of the beginnings of the history of Paris. 'The dialectical image' for Benjamin (1999a, p. 474) is 'the primal phenomenon of history.' Lutetia is the primal phenomenon of the history of Paris. Of course, Lutetia can be and has been forgotten by the majority of the residents of Paris, but that does not mean, as Benjamin says, that 'it does not extend into the present. On the contrary: it is present by virtue of its very oblivion,' as Benjamin (1999b, p. 809) said of the hetaeric stage and Bachofen's conception of matrifocal societies (Bachofen, 1967). Lutetia extends into the present and is present by virtue of its very oblivion in Paris today and in it and other marsh metropolises and swamp cities. Benjamin's comment is a reminder of the motto for Paris ('it floats but does not sink'), as well as a fitting motto for *Cities and Wetlands* (Giblett, 2016) and for the return of the repressed in nature and culture.

The modern city not only takes place in a place in space but also occurs in a moment in time, in a period of history. The hell of the modern city is not only found in the city itself, in its dark and dirty places, but also in the hell of the modern, in its hot and foetid events. Modernity for Benjamin (1999a, p. 842, see also 1999a, p. 544) is 'the time of hell.' Why? Because 'the eternity of hell' is constituted by the fact 'that which is newest . . . remains, in every respect, the same.' The modern is cut off from pre-history and is constituted by its severance from pre-history. Most residents of all the cities discussed in *Cities and Wetlands* (Giblett, 2016) have no knowledge of the pre-history of the city in which they live. Benjamin (1999a, p. 544) earlier on the same page defined the 'modern' as 'the new in the context of what has always already been there.' The modern city in general and Paris in particular are new in the context of the swamp that has always already been there.

The temple of Lutetia of the grotesque lower earthly (Giblett, 1996) and bodily strata (as Bakhtin (1984) called it), of the monstrous feminine *was* the swamp, whereas the arcades are temples of the God of greedy capitalism, of the capital(ist) city, of new, modern, monumental Paris. The arcades, or interior passageways, of Paris in the nineteenth century for Benjamin (1999a, pp. 37 and 546), are 'the hollow mold form which the image of "modernity" was cast' and 'temples of commodity capital.' The city/monster displaces and takes over the place of the swamp/monster. Benjamin cites an illustrated guide to Paris of 1852 in which 'an arcade is a city, even a world, in miniature' (cited by Benjamin, 1973a, pp. 37 and 158). An arcade is a miniature city and world of commodity capitalism that displaces and replaces the macro-earth and economy of swamps and marshes that nevertheless leave traces in the history of, and metaphors for the city of Paris, and its sewers, and its pre-history as marsh.

Note

1 For further discussion of aura and trace in relation to photography see Giblett and Tolonen, (2012, chapter 1).

References

Adorno, T. and W. Benjamin. 1999. *The Complete Correspondence: 1928–1940.* H. Lonitz, ed. N. Walker, trans. Cambridge, MA: Harvard University Press.

Aragon, L. 1994. *Paris Peasant.* S. Taylor, trans. Boston: Exact Change. First published 1926.

Bachofen, J. J. 1967. *Myth, Religion and Mother Right: Selected Writings*. R. Manheim, trans. Princeton: Princeton University Press.

Bakhtin, M. 1984. *Rabelais and His World*. H. Iswolsky, trans. Bloomington: Indiana University Press.

Benjamin, W. 1973a. *Charles Baudelaire: A Lyric Poet in the Era of High Capitalism*. H. Zohn, trans. London: Verso.

Benjamin, W. 1973b. *Illuminations*. H. Zohn, trans. London: Fontana.

Benjamin, W. 1979. *One-way Street and Other Writings*. E. Jephcott and K. Shorter, trans. London: NLB.

Benjamin, W. 1996. *Selected Writings, Volume 1: 1913–1926*. M. Bullock and M. Jennings, eds. Cambridge, MA: Belknap Press of Harvard University Press.

Benjamin, W. 1999a. *The Arcades Project*. H. Eiland, trans. Cambridge, MA: Belknap Press of Harvard University Press.

Benjamin, W. 1999b. *Selected Writings, Volume 2: 1927–1934*. M. Jennings, H. Eiland and G. Smiths, eds. Cambridge, MA: Belknap Press of Harvard University Press.

Benjamin, W. 2002. *Selected Writings, Volume 3: 1935–1938*. H. Eiland and M. Jennings, eds. Cambridge, MA: The Belknap Press of Harvard University Press.

Benjamin, W. 2003. *Selected Writings: Volume 4, 1938–1940*. H. Eiland and M. Jennings, eds. E. Jephcott and others, trans. Cambridge, MA: The Belknap Press of Harvard University Press.

Benjamin, W. 2014. *Radio Benjamin*. J. Lutes and others, trans. L. Rosenthal, ed. London: Verso.

Cohen, M. 1993. *Profane Illumination: Walter Benjamin and the Paris of Surrealist Revolution*. Berkeley: University of California Press.

Deleuze, G. 1989. *Masochism: Coldness and Cruelty*. New York: Zone Books.

Dodds, J. 2011. *Psychoanalysis and Ecology at the Edge of Chaos: Complexity Theory, Deleuze/Guattari and Psychoanalysis for a Climate in Crisis*. London: Routledge.

Ebbatson, R. 2013. *Landscape and Literature 1830-1914: Nature, Text, Aura*. Basingstoke: Palgrave Macmillan.

Eiland, H. and M. Jennings. 2014. *Walter Benjamin: A Critical Life*. Cambridge, MA: Belknap Press of Harvard University Press.

Giblett, R. 1996. *Postmodern Wetlands: Culture, History, Ecology*. Edinburgh: Edinburgh University Press.

Giblett, R. 2008a. *The Body of Nature and Culture*. Basingstoke: Palgrave Macmillan.

Giblett, R. 2008b. *Sublime Communication Technologies*. Basingstoke: Palgrave Macmillan.

Giblett, R. 2009. *Landscapes of Culture and Nature*. Basingstoke: Palgrave Macmillan.

Giblett, R. 2011. *People and Places of Nature and Culture*. Bristol: Intellect Books.

Giblett, R. 2016. *Cities and Wetlands: The Return of the Repressed in Nature and Culture*. London: Bloomsbury.

Giblett, R. 2018. *Environmental Humanities and Theologies: Ecoculture, Literature and the Bible*. London: Routledge.

Giblett, R. and J. Tolonen. 2012. *Photography and Landscape*. Bristol: Intellect Books.

Gunning, T. 1986. The Cinema of Attraction. *Wide Angle*, 8(3–4), pp. 63–70.

Gunning, T. 1989. An Aesthetic of Astonishment: Early Film and the (In)credulous Spectator. *Art and Text*, 34, pp. 31–45.

Johnson, L. 1988. *The Unseen Voice: A Cultural Study of Early Australian Radio*. London: Routledge.

Mules, W. 2014. *With Nature: Nature Philosophy as Poetics Through Schelling, Heidegger, Benjamin and Nancy*. Bristol: Intellect Books.

Neale, S. 1985. *Cinema and Technology: Image, Sound, Colour*. London: Macmillan.

Pike, D. L. 2007. *Metropolis on the Styx: The Underworlds of Modern Urban Culture, 1800–2001*. Ithaca, NY: Cornell University Press.

Scholem, G. 2012. *On Jews and Judaism in Crisis: Selected Essays*. W. Dannhauser, trans. Philadelphia: Paul Dry.

Solnit, R. 2000. *Wanderlust: A History of Walking*. London: Penguin.

Virilio, P. 1989. *War and Cinema: The Logistics of Perception*. P. Camiller, trans. London: Verso.

Virilio, P. 1994. *The Vision Machine*. J. Rose, trans. Bloomington: Indiana University Press.

Virilio, P. 2000. *The Information Bomb*. C. Turner, trans. London: Verso.

Wolin, R. 1982. *Walter Benjamin: An Aesthetic of Redemption*. New York: Columbia University Press.

6 The uncanny cyborg

By mimicking human actions, gestures and behaviour dolls and automata were particularly uncanny for Freud (2003, p. 135) as they evoked and elicited both horror and fascination because they are both like humans and different from humans. In the age of communication and information technologies the boundary and distinction between human and machine is becoming increasingly blurred with the cyborg, the cybernetic organism. This chapter considers the cinema-goer as the first cyborg and the cyborg in conjunction with information technologies. The uncanny cyborg is also a persistent topic in visual arts and culture as presented and portrayed in a landmark exhibition in Vancouver 2001 and resulting catalogue (Grenville, 2001). Human beings connected to and by information technologies are cyborgs, but whether they are (also) symbionts on a symbiotic planet ushering in the symbiocene is another question. This chapter makes the plea for being and doing so.

Film, for Benjamin (1986, p. 55) writing before the advent of television, was 'one of the most advanced machines for the imperialist domination of the masses.' Yet the cinematic machine is not only outside 'the masses' dominating them (us?) from without, but also inside our own heads dominating them/us from within. We are complicit with the machine and collude in our/its domination: 'The machine is us,' as Haraway (1985, p. 99) says. The cinematic institution for Metz (1982, p. 7) 'is not just the cinema industry . . ., it is also the mental machinery – another industry – which spectators 'accustomed to the cinema' have internalised historically and which has adapted them to the consumption of films.' 'The mental machinery' of cinema, as Neale (1985, p. 1) puts it, is 'an apparatus for the production of meanings and pleasures, and as such involves aesthetic strategies and psychological processes.' There is no distinction between the machine and us insofar as we are constituted as mental beings. The machine is us mentally. Cinema is a sublime communication technology of mind-body dualism which in Carey's (1989, p. 206) words 'locates vital energy in the realm of the mind' and not, by implication, in the realm of the body.

There is a distinction between the machine and us insofar as we are bodily beings. The machine replaces physical work and activity, but it is a mental entity before it is a physical one, an idea before an invention, and it is internalized in our minds rather than in our bodies. This internalization has become naturalized over a hundred years of cinema-going, but it was entirely new for the first cinema-goers and cinematographers, or more precisely 'camera-eye men.' Dziga Vertov (1984, pp. 14–15), 'the man with a movie camera,' is usually seen as the apostle and evangelist (and no apologist) for the camera as eye and for 'the use of the camera as a kino-eye, more perfect than the human eye.' Not that the human eye is imperfect, just that the camera is more perfect. Vertov (1984, p. 17) was also the polemicist and proponent of the body as vehicle for the camera. Both camera as perfected eye and body as camera-vehicle come together in the boast that 'I am kino-eye, I am a mechanical eye. I, a machine, show you the world as only I can see it.'

Vertov (1984, p. 6) elaborated the implications of this mind-machine-eye-camera-body combo in his 'We: Variant of a Manifesto,' in which he announced that 'we call ourselves *kinoks* ("cinema-eye men") as opposed to "cinematographers."' Kinoks for Vertov (1984, p. 7) 'satisfy 'man's . . . desire for kinship with the machine' as 'we bring people into closer kinship with machines' (Vertov, 1984, p. 8). 'Man' and machine are constituted as members of the same family, or at least the machine is a long lost relative with which 'man' wants to establish his filial relations. In order to overcome the alienation of 'man' and machine, the kinok wants to acknowledge that the machine is related to and produced by 'man.' And he wants to enact that kinship in the act of filming.

Yet in the process the machine is anthropomorphized, or at least animated, and rhapsodized. To become machine, 'man' has to animate machine. Vertov (1984, p. 9) gives an 'hurrah for the poetry of machines, propelled and driving the poetry of levers, wheels, and wings of steel, the iron cry of movements, the blinding grimaces of red-hot streams.' The machine is figured as an animated body (in two senses, animal body and moving body, not static body of ore) that emits terrifying screams and burns out the eyes. The machine is an instrument of the law of the Father that operates through terror and the threat of castrating the phallic power of sight.

The machine for Vertov (1984, p. 8) even has a soul and the poetry of machines is in the business of 'revealing the machine's soul.' Fulfilling 'man's desire for kinship with the machine and its soul' means transforming 'man' into machine, albeit an animated, rhapsodized, even sublimated, machine. For Vertov (1984, p. 8; emphases in original) '*The new man*, free of unwieldiness and clumsiness, will have the light, precise movements of machines.' Kinoks sublimate the heavy and solid, cumbresome and clumsy human body into the light and ethereal, dexterous and precise movements of the machine. For Vertov (1984, p. 8; emphases in original) and his fellow kinoks '*our path leads through the poetry of machines, from the bungling citizen to the perfect electric man*.'

Unlike Whitman in his *Leaves of Grass* who merely *sang* the body electric (see Benthall, 1976, pp. 13 and 163–171), Vertov in his *Man with a Movie Camera* and as the man with a movie camera *made* the body electric just as the cinema in general electrified the human body (see Christie, 1994, pp. 64–87). Photography enacted the Promethean desire to snatch fire from the gods in the heavens and bring it down to humans on earth; cinema achieved that desire in what Christie (1994, p. 65) calls, using Mary Shelley's subtitle of *Frankenstein*, 'the Modern Prometheus.' Electricity is the modern fire; cinema is one of its vehicles for coming down to earth from the heavens (radio is another; see Giblett, 2008b, chapter 7); the cinematographer (and the star as we saw in the previous chapter) are the body electric.

Similarly, electricity for Villiers de l'Isle Adam (1981, p. 149; see also p. 77) was 'the spark bequeathed by Prometheus.' In his novel of the 1880s, *Eve of the Future Eden*, a fictionalized Edison 'with the sublime aid of light' creates what is variously described as 'a magneto-electrical entity,' 'a new electro-human creature,' 'an electro-magnetic creature' or simply 'Andreid' (de l'Isle-Adam, 1981, pp. 67, 72, 113 and 180), the first fictional Android, called Hadaly (meaning the Ideal). In the whirring armatures and phonographic speech of 'the magnetic-metallic organism of Hadaly' (de l'Isle-Adam, 1981, p. 92), Edison claims he will 'bring illusion down to earth . . . and force the Ideal to show itself for the first time to your senses, palpable, audible, and materialised' (de l'Isle-Adam, 1981, pp. 72–73).

What Shelley and l'Isle-Adam only dreamed and wrote about cinema was to realize. Cinema brought dead matter to life on the screen. Hadaly as 'a regal vision machine, almost a creature, a dazzling simulacrum' (de l'Isle-Adam, 1981, p. 97) was the forerunner of the cinematic cathedral and televisual cathode-ray 'vision machine' (Virilio, 1994) of cinema and television. She is a dead living star. Hadaly, the Andreid, 'knows no life, no disease, no death. She is above all imperfections and all servitudes' orbiting in the sublime company of heavenly bodies where 'she keeps her ethereal beauty' pure and unsullied, like the dead living star on the silver screen (de l'Isle-Adam, 1981, p. 177).

The modern communications technologies of photography and cinema employed the light of the sun, the fire in the sky. Cinema later employed the artificial light of the modern fire of electricity to project film on to a screen. Modern industrial technologies unbound Prometheus (see Landes, 1969, especially pp. 24 and 284) to release the power of heat in thermodynamic technologies (steam and internal combustion engines), and of light and electricity in communication technologies (photography, cinema, telegraphy, radio and television). All of them entailed what Landes (1969, p. 24) calls 'mastery over nature' and 'mastery of the environment' rather than mutuality with them (see Giblett, 2011).

Electricity is the connection between the body of 'man' with its nerves and the cinema with its light machines and displays. And its nerves. For Vertov (1984, p. 8), 'Cinema's unstrung nerves need a rigorous system of precise movement.' It needs nerves of steel, nerves of electrical impulses. Cinema needs a grid of wires to circulate power and energy and to control electrical circulation through command, control and feedback. Cinema needs and uses what Vertov (1984, p. 7) called 'electricity's unerring ways.' Electricity constitutes cinema as machine and the human body as electrical machine. Electricity becomes the model for cybernetic control of the human body by way of the nervous system.

With cinema, human beings become a cybernetic organism, a cyborg. The first cyborg was the cinematographer and cinema-goer. Previously in the physiology of Leonardo da Vinci and René Descartes the human body was a machine; in Vertov's physiology the human body becomes an electric motor.

Previously for Leonardo da Vinci the earth also was a machine; with Vertov the whole earth is an electric motor. For Vertov (1984, p. 8):

> Openly recognizing the rhythm of machines, the delight [*sic!*] of mechanical labour, the perception of the beauty of chemical processes, WE [*sic*] sing of earthquakes, we compose film epics of electric power plants and flame, we delight in the movements of comets and meteors and the gestures of searchlights that dazzle the stars.

Searchlights, as Bailes (1980, p. 67) puts it, 'dramatically symbolized the new technology [of] electrical illumination' and were first used in war during the Boer War (as he goes on to relate). They were also used in the lead up to the WWII when Albert Speer created Zepplinfield composed of 'walls of light' using 150 searchlights (see Virilio, 1989, p. 78). Zepplinfield and cinema both produce 'photo-murals.' The play of light is fleeting, evanescent but fixed in virtual walls that immobilize the spectator. Cinema is a sublime communication technology as it deals with creative, large-scale events and transforms them into soaring and transcending walls of light, into virtual walls.

'Man' becomes a cybernetic organism as the culmination of the transition from the body AS machine (in modern Western medicine, Fascism and sport) where the machine is a figure or metaphor for the body (see Giblett, 2008a) to the body AND machine (in technologies, especially communication technologies). From the machine-body of modern Western medicine, through the Fascist body of the war-machine, to the body-machine of the civilian-soldier, 'Man' becomes, in a word, a cyborg. The term 'refers to [a?] cybernetic organism, a self-regulating human-machine system' (Featherstone and Burrows, 1995, p. 1). Cybernetics is the study and use of command and control systems in machines and organisms (see Wiener, 1989, pp. 15–17). As command and control involve feedback, cybernetics is more precisely 'the science of automatic, self-regulated control' (Levidow and Robins, 1989a, p. 8).

The term 'cybernetics' comes from the Greek word for 'the steersman' (see Wiener, 1989, p. 15) and from 'the governor' of nineteenth-century steam machines and thermodynamics (see Tomas, 1995, p. 23). Cybernetics can thus be considered as an extension of what Foucault (1979, pp. 5–21, 1991, pp. 87–104) calls governmentality. He used this term to describe all the ways and means by which the lives of populations are policed by state apparatuses. It can be extended to apply to the functioning of the body self-regulated by communication technologies. It also applies to the functioning of the body policed by transnational corporations, the successor to the state as the dominant institution of our hypermodern age. Cybernetics can also be considered as an extension of the internalization of the panoptic principle of self-surveillance (see Foucault, 1977) through the use of communication technologies (see Giblett, 2008b).

Yet the cyborg is not only a product of science but also a creature of science fiction. A cyborg for Haraway (1985, p. 65) is 'a cybernetic organism, a hybrid of machine and organism, a creature of social reality as well as a creature of fiction.' The cyborg for her is a matter of lived experience and of science fiction 'full of cyborgs – creatures simultaneously animal and machine, who populate worlds ambiguously natural and crafted' (Haraway, 1985, p. 66). On the side of social reality and lived experience, modern medicine for Haraway (1985, p. 66) is 'also full of cyborgs, of couplings between organism and machine' such as IV drips, heart-rate monitors and pacemakers. On this side too 'modern war is a cyborg orgy, coded by C^3I, command-control-communication-intelligence.'

Cybernetics began in Wiener's work on predictive gun-sights during the Second World War. C^3I reached highly visible manifestation in the Gulf War of 1991 where satellites and computers were used with deadly accuracy in aiming smart bombs. Cyborgs abound not only in war, but also in non-war, not only in the military but also in civilian life. For Gray (1989, p. 67), one of Haraway's students, 'In Europe and North America [in Australia, New Zealand, Japan] we are cyborgs in many ways already: medical implants in our bodies, mechanical connections close at hand, products from around the world in every corner of our lives.' Globalization makes us into cyborgs not only by bringing products to us via transportation but also by bringing TV and computer programs to us via communication. The cyborg is not so much a matter of hardware, but of software, of programs and coding. It is also a matter of software and wetware coming together, and of wetware being programmed and coded.

The cyborg is cultural and natural; the cyborg is a creature, and creation, of culture and nature. More to the point, the cyborg is the creature of the *opposition* of culture to nature. The human body is the creation of both culture and nature. Capitalist modernity instituted a binary opposition and a hierarchical relationship between culture and nature (see Balsamo, 1995, pp. 215 and 217; and Giblett, 2011, chapter 2). Culture was privileged over nature, and culture was pitted against nature in such a way that the former would inevitably win and the latter lose. Capitalist modernity ascribed the body to nature and the mind to culture. It also figured the body and nature as machine produced by culture and the mind, and so prefigured the cyborg.

The cyborg is a creature that crosses the boundary of the capitalist opposition of culture to nature and that hybridizes the two (rather than deconstructs or decolonizes this opposition). Rather than 'becoming cultural bodies' and already being them as Dyens (2001, p. 2; see also p. 19) argues, we have always been cultural bodies (see Giblett, 2008a). The cyborg is different in degree rather than in kind from, say, the gunner and cinema-goer, and they in turn from, say, the hunter and the cave painter. The difference between them lies in the different 'cultures of natures' they enact, between the first nature of first peoples and cultures, of the cave painter; the second nature of agriculture, mining and cities; the third nature of modernity, of the cinema-goer; and the fourth nature of hypermodernity, of the cyborg, though the cyborg in fact is prefigured by the cyborg in fiction of high modernity (for the 'cultures of natures,' see Giblett, 2011, especially chapter 1).

The cyborg, as Featherstone and Burrows (1995, p. 2) define it, is 'a human-machine hybrid.' It tries to combine the best of both worlds, of both humanity and machinery (but also produces the worst of all possible worlds). Rather than a liminal or transgressive or even monstrous figure, the hybrid is the norm (Latour cited by Lykke, 2000, p. 77), especially when it comes to gender for the body, nature and the machine have been gendered as female/feminine whereas the mind, culture and the machinist (inventor, scientist, technician, mechanic) have been gendered as male/masculine. The cyborg is a hybrid that crosses the culture/nature, mind/body and male/female divides. A cyborg may be feminized as with the female cyborgs of film and novel such as Maria in Fritz Lang's *Metropolis* of 1926 or 'the magnetic-metallic organism' of Hadaly in Villiers de l'Isle Adam's *L'Eve future* written in 1886 (de l'Isle Adam, 1981, p. 92; see also Balsamo, 2000, pp. 149–150). Or it may be masculinized as with the male cyborgs of the *Terminator* films.

Cyborgs, Balsamo (2000, pp. 150 and 155) argues, 'fascinate us by technologically refashioning human difference . . . because they are not like us, and yet just like us.' Cyborgs are uncanny, just as automatons were for Freud. As artefacts that 'come alive' or mimic life they fascinate us with their mechanical reproduction of human aptitudes and they terrify us with their resemblance to human mind and bodies. They cross the human-machine divide and bring the news that the difference is not all that great. The cyborg is the uncanny flipside to sublime communication technologies.

Sublime communication technologies are an integral part of life today. Just as we do not get much choice about whether or not we use them, so we do not get to choose whether or not we are cyborgs. 'We are cyborgs' whether we like it not as Haraway (1985, p. 66; see also Featherstone and Burrows, 1995, pp. 3 and 10) puts it. 'We are not becoming cyborgs,' as Dyens (2001, pp. 8 and 88) puts it, as 'we ourselves have become cyborgs.' Every time we use the phone, the computer, the ATM, have an injection or medical test, pop a pill, drive a car, take a bus, watch TV, listen to the radio, see a film, play a video game, surf the Net, send an email, we are cyborgs. Our bodies have entered into some sort of dependence on, and enhancement by, a machine, principally a communication technology. We have ceased to be our own body exclusively and we have become a body-machine hybrid, a cyborg. 'The machine is us,' Haraway (1985, p. 99) concludes. The machine is not something outside us that we control and manipulate, and then stow away or dispose of, but something that becomes us. As Dyens (2001, p. 32) concurs, 'We are machines and the machine is within us.'

Yet the cyborg is more precisely the interface with an electronic machine rather than with an electromagnetic machine or an internal combustion machine. The body as machine, as made up of parts, is as old as the Renaissance; the nervous system as telegraphic control system as old as telegraphy; the body electric is as old as the discovery of electricity and its use in electric motors. The body electronic, what Wiener (cited by Tomas, 1995, p. 23) calls 'the body as an electronic system,' is the creature of what Tomas (1995, p. 23) calls 'the age of communication and control.' Not only, as for Cubitt (1998, p. 50) 'the cyborg is alive, and working for RTZ,' but also anybody who works for any corporation or university for that matter, just like Cubitt, is a cyborg.

Yet the cyborg is not merely used in war as a weapon, but is the child of war. Whether it is the legitimate or illegitimate child of war is another question. Haraway (1985, p. 68) concedes that 'the main trouble with cyborgs, of course, is that they are the illegitimate offspring of militarism and patriarchal capitalism.' By arguing that the cyborg is *il*legitimate Haraway is then able to use it as a figure of irony and resistance against militarism and patriarchal capitalism. Yet the cyborg is the *legitimate* child of militarism and patriarchal capitalism as they are married and as they gave birth to modern machines and bodies in industry and war from the first factories, the first rapid firing guns, and the first use of communication and transport technologies, such as the railway and telegraphy, in both war and industry (see Giblett, 2008b).

In fact, militarism and patriarchal capitalism were married and gave birth to war-machines and to the soldier from their very beginnings. The cyborg is the youngest offspring of this marriage that in its latest twist gave birth to the soldier-civilian in the increasing militarization of civilian life. 'All of us are already civilian soldiers, without knowing it,' says Paul Virilio. The cyborg is not only a hybrid of machine and organism, but also of soldier and civilian, or perhaps more precisely it embodies or empanels the colonization of the civilian by the soldier, the grafting of the soldier on to the civilian. The cyborg is a creation of modern militarism and militaristic modernity. The cyborg is a creature, figure and product of the militarization of civilian life. As such its genealogy can be traced at least back to Fascism that aestheticized civilian life with military style. Cybernetics was a military creation in a long line of communication technologies with beginnings in the military (see Giblett, 2008b). The cyborg is the *legitimate* child of militarism and patriarchal capitalism.

The point of establishing the *il*legitimacy of the cyborg for Haraway is that it provides leverage in resisting militarism and patriarchal capitalism by virtue of being illegitimate. In particular, it provides the irony that Haraway and so many other post-modern theorists see as necessary for resistance. Yet given the dubious parentage of the cyborg, but still acknowledging that we are cyborgs, irony and resistance lie more effectively in the grotesque and the monstrous, in recoupling the cyborg with the what Haraway (1995) later called the symbiont. We are cyborgs, but we are also symbionts who are in symbiosis with the oxygen-producing plants of this planet, with other animals, with bodies of water and with other bodies of earth. Whether this symbiosis is mutual and sustainable or masterly and parasitic is another matter (see Giblett, 2011, especially chapter 12).

The boundaries between ourselves as symbionts and as cyborgs are being blurred as cybernetics is reshaping our bodies. Communications technologies and biotechnologies are for Haraway (1985, p. 82) 'the crucial tools recrafting our bodies.' These technologies, and the 'communication sciences and modern biologies' associated with them, are constructed for Haraway (1985, p. 83; emphases in original) by 'a common move – *the translation of the world into a problem of coding.*' Although undoubtedly both communication and biological sciences and technologies *do* translate 'the world' into a problem of coding, the question remains *should* they, and indeed *can* they without reducing the world to a set of numbers and mathematical formulae and sets of calculations, to digital data, to binary oppositions. And thereby killing it. We not only murder to dissect as Wordsworth said, but also murder to decode and encode.

Communication and cultural students and semioticians may respond positively to the mention, and notion, of code, but this is a dubious response that takes us back into the C^3I (command-control-communication-intelligence) composite of militarism and patriarchal capitalism. To encode (and decode) is to reduce communication to the transportation of messages (see Giblett, 2008b). Even the seemingly innocuous communication model of Shannon and Weaver has military overtones not only because of Weaver's successful application of mathematical and statistical techniques in deciphering enemy messages and any simply authorial intentionalism that would ascribe a causal relationship between his employment and his theory (see Hutchins, 1999, pp. 5–6), but also because encoding and decoding messages is the stock-in-trade of war.

Not only coding bodies *as* machines as in modern Western medicine, but also coupling bodies *to* machines as in modern Western technologies has effects on both bodies and machines, and changes their meaning, and nature. Evelyn Fox Keller (1994, p. 316) wonders, can it be 'any surprise that, in modelling organisms and machines, each upon the other, not only do organisms and machines come increasingly to resemble each other but also that . . . the meaning of both terms undergoes some rather critical changes?' Foremost amongst these changes for Keller (1994, p. 321) is the way in which 'the body of modern biology, like the DNA molecule – and like, too, the modern corporate or political body – has become just another part of an informational network, now machine, now message, always already ready for exchange, each for the other.' To code the body and the self is to make it communicable semiotically and transportable virtually.

When Haraway (1985, p. 82) argues that the cyborg self is 'the self feminists must code,' she is advocating a militarist and patriarchal capitalist self rather than a self and a body that cannot be reduced to a code. Rather than a cyborg self, a playful, grotesque, monstrous body resists all attempts to encode and decode it. To figure the body and self as land is to connect it to a place and a community that cannot be communicated or transported without killing it. Similarly, to code the earth as Gaia as Lovelock has done is to code it as cyborg, as Haraway (1995, pp. xiii–xiv; see also Dyens, 2001, p. 49) points out. It is not to live with the earth as living body, as goddess. It is merely to use Gaia as a metaphor whereas the earth is our mother beyond mere metaphor (see Giblett, 2011, chapter 9).

Yet the earth, like the body, has not only been encoded but also subjected to, and represented by, visual technologies. Just as the body has been encoded in DNA and in the Human Genome Project and investigated by x-rays, cat-scans and ultrasound, so has the earth been photographed from outer space and constituted as scientific object, as 'the environment' (see Haraway, 2000, p. 222). As discourses constitute their objects, so for Haraway (2000, p. 222) 'both the whole earth and the foetus owe their existence as public objects to visualizing technologies. . . . The global foetus and the spherical whole earth both exist because of, and inside of, technoscientific visual culture.' The biosphere ceases to have an existence, let alone life, outside of visualizing technologies. As some aspects of the ecosphere, such as the electromagnetosphere and global extraterrestrial space, are invisible, they are not acknowledged, let alone respected, as part of the earth household.

In a peculiarly hypermodern twist the whole earth photographed from outer space and the foetus photographed in inner space come to stand for the sign itself in what Haraway (2000, p. 225) goes on to call 'ordinary magico-secular transubstantiation' in which the image or symbol is transformed into the body and blood of the foetus and the earth. Instead of the sign *representing* the thing as in modernity, and the sign *being* the thing as in pre- or non-modernity (see Giblett, 2011), 'the sign *becomes* the thing itself' (Haraway, 2000, p. 225 (my emphasis)). The hypermodern moment is not merely signs disconnected from things, but signs become the things themselves (see Giblett, 2008b) when the thing itself (foetus and globe) is not accessible to ordinary human vision either because it is hidden from view or too big to be seen.

By being figured as machine, by being coded as a message, by being scanned and x-rayed, the body has ceased to be the ground of our being and the ultimate matter of our life – though it is still the ultimate matter of our death. Bodies, according to Virilio (2000, p. 11), 'are no longer the ultimate matter, our skin the final frontier, our consciences the training ground for a world turned on its head. The new limits are now to be found *beyond*, in otherwise transcendent realms.' The ultimate matter is light beyond whose speed nothing can travel. Bodies are constituted as a frontier to be crossed, not a home to dwell in, but they are still a constraint, a limit-case. Death is the ultimate limit, and matter, dead matter. The body is living matter, or at least the living body is. Or the body is living or dead. A dead body is still a body. The body can be either alive or dead. What is the body? The set of biological and physical processes and structures that lives and dies.

Just as the machine-body gives rise to a disembodied self in modern Western medicine and sport, the cyborg for Haraway (1985, p. 82) gives rise to what she calls 'a kind of disassembled and reassembled, post-modern collective and personal self.' This is a good self, whereas, in strong contrast to Haraway, Robins and Levidow (1995a, p. 119, 1995b, pp. 105–106; see also Levidow and Robins, 1989b, p. 172) characterize the cyborg self in the following terms:

> Through a paranoid rationality, expressed in the machine-like self, we combine an omnipotent phantasy of self-control with fear and aggression directed against the emotional and bodily limitations of mere mortals.

> Through regression to a phantasy of infantile omnipotence, we deny our dependency upon nature, upon our own nature, upon the 'bloody mess' of organic nature.

In what they go on to call 'a sublimatory, compensatory control' (Levidow and Robins, 1989b, p. 172) the cyborg self thereby 'transcend[s] human limitations' (Levidow and Robins, 1989b, p. 168), especially the earthly and natural limits and prison of the body, into the sublime realm of heavenly bodies. As the sublime for Kant is the category by which we deny our dependence upon nature, the cyborg self encapsulates and embodies the sublime (see Giblett, 1996, chapter 2). The cyborg self is sublime. It is a self linked to sublime communication technologies (see Giblett, 2008b, chapter 9) like a life-support system.

Robins' and Levidow's cyborg self is akin to Haraway's (1995, p. xi) later characterization of what she calls 'the Terminators' drawing principally on the second film: 'The jelled-metal, shape-shifting, cyber-enhanced warriors fighting in the stripped terrain landscapes and extraterrestrial landscapes of a terrible future.' Even ordinary cyborgs like ourselves are 'cyber-enhanced' so in dealing with the difference between the cyborg and the Terminator we are dealing with a difference in degree rather than kind. The term 'cyborg,' Haraway (1995, p. xv) goes on to point out, was coined by Clynes and Kline in 1960 to refer to 'the enhanced man who could survive in extraterrestrial environments.' There is not much difference between the cyborg and Terminator here. Nor much difference between astronauts and cybernauts like ourselves who survive in the extraterrestrial environment of the electromagnetic spectrum, the ether of telecommunications, and 'off-world' in the global network of cities connected by an umbilical cord, like an astronaut, to mother-earth-ship (see Giblett, 2011 cover photograph and chapter 12).

The cyborg and the Terminator are not that much different. Perhaps their difference lies in gender, in their gender. The Terminator is the bad boy of post-modern culture, whereas the cyborg for Haraway is the bad girl or good bitch of socialist feminism, the cyberfeminist. Haraway (1985, p. 101) would rather be a cyborg than a goddess, but Lykke (2000, pp. 75, 82 and 85) would like to be a cyborg *and* a goddess, a 'cybergoddess.' The Terminator for Haraway (1995, p. xv) is 'the sign of the beast on the face of post-modern culture, the sign of the Sacred Image of the Same. . . . The Terminator is the self-sufficient, self-generated Tool in all of its infinite but self-identical variations.' The Terminator is not simply the Fascist machine-body but the Fascist machine-tool-body that manufactures machines. The cyborg is the legitimate child of the marriage between militarism and patriarchal capitalism.

Whereas we can refuse to be Terminators, we are already cyborgs. Haraway (1995, p. xix) concludes by suggesting that:

> I do not think that most people who live on earth now have the choice not to live inside of, and not be shaped by, the fiercely material and imaginative apparatuses for making 'us' cyborgs and making our homes into places

> mapped within the space of titanic globalisations in a direct line of descent from the cybernetic Gaia seen from NASA's fabulous eyes.

Most people who live on earth don't have access to electrical power, phone lines, computers and Western medicine so they are not cyborgs though they may not have much choice about whether or not they are targeted to become cyborgs. As for those of us who have access to these things, we may not have any choice about being cyborgs but we do have choice about being Terminators. We can choose to be either 'Cyborgs for Earthly Survival' as Haraway (1995, p. xix) puts it or 'Terminators for Earthly Destruction.' We may not have any choice about living on earth but we do have choice about whether we regard the earth as 'cybernetic Gaia,' as 'a dynamic, self-regulating, homeostatic system' (yuk!) as Haraway (1995, p. xiii) puts it, or as living, uncodable home, as grotesque and monstrous body, fecund and procreative, life-giving and death-dealing. We are cyborgs and symbionts, but we have to choose whether we will also be goddesses and symbionts for earthly survival or Terminators for earthly destruction.

In her later writing about the cyborg, Haraway complements the figure of cyborg with that of the symbiont in order to foreground the organic component of the cyborg. The cyborg is a composite cybernetic organism, but the organism is often overlooked in the emphasis on the cybernetic. The shiny hardware and calculating software are often emphasized to the detriment of the messy wetware that helps to make up the cyborg, that makes it possible for the cyborg to survive and function. Without the organism, there is no cyborg. In coupling the symbiont with the cyborg Haraway draws on the work of Lynn Margulis. For nearly 30 years Margulis (1998, p. 33; see also Warming, 1909, pp. 83–95) was the most cogent proponent of the notion of symbiosis, 'the term coined by the German botanist Anton de Bary in 1873.' Margulis (1971, p. 49; see also 1981, p. 161) defines symbiosis as 'the living together of two or more organisms in close association. To exclude the many kinds of parasitic relationships known in nature, the term is often restricted to associations that are of mutual advantage to the partners.' As Robert Frodeman (1992, p. 319) argues, 'We are in symbiotic relation to the oxygen-producing plants of the world' – whether we like it or not. Every breath of air we take re-affirms this symbiosis. As Margulis (1998, p. 5) puts it, 'We are symbionts on a symbiotic planet.' Whether this is a parasitic and destructive relationship for the host, or mutually beneficial symbiosis for guest and host is another question (see Giblett, 2011, chapter 12).

References

Bailes, H. 1980. Military Aspects of the War. In: *The South African War: The Anglo-Boer War 1899–1902.* P. Warwick, ed. London: Longman, pp. 65–102.

Balsamo, A. 1995. Forms of Technological Embodiment: Reading the Body in Contemporary Culture. In: *Cyberspace/Cyberbodies/Cyberpunk: Cultures of Technological Embodiment.* M. Featherstone and R. Burrows, eds. London: Sage, pp. 215–237.

Balsamo, A. 2000. Reading Cyborgs Writing Feminism. In: *The Gendered Cyborg: A Reader.* G. Kirkup, L. James, K. Woodward and F. Hovenden, eds. London: Routledge, pp. 148–158.

Benjamin, W. 1986. *Moscow Diary.* R. Sieburth, trans. G. Smith, ed. Cambridge, MA: Harvard University Press.

Benthall, J. 1976. *The Body Electric: Patterns of Western Industrial Culture.* London: Thames & Hudson.

Carey, J. 1989. Technology and Ideology: The Case of the Telegraph. In: *Communication as Culture: Essays on Media and Society.* Boston: Unwin Hyman, pp. 201–230.

Christie, I. 1994. *The Last Machine: Early Cinema and the Birth of the Modern World.* London: BBC Educational Developments.

Cubitt, S. 1998. *Digital Aesthetics.* London: Sage.

Dyens, O. 2001. *Metal and Flesh: The Evolution of Man: Technology Takes Over.* E. Bibbee and O. Dyens, trans. Cambridge, MA: The MIT Press.

Featherstone, M. and R. Burrows. 1995. Cultures of Technological Embodiment: An Introduction. In: *Cyberspace/Cyberbodies/Cyberpunk: Cultures of Technological Embodiment.* M. Featherstone and R. Burrows, eds. London: Sage, pp. 1–17.

Frodeman, R. 1992. Radical Environmentalism and the Political Roots of Postmodernism: Differences That Make a Difference. *Environmental Ethics*, 14(4), pp. 307–319.

Foucault, M. 1977. *Discipline and Punish: The Birth of the Prison.* A. Sheridan, trans. London: Penguin.

Foucault, M. 1979. On Governmentality. *I & C*, 6, pp. 5–22.

Foucault, M. 1991. Governmentality. In: *The Foucault Effect: Studies in Governmentality.* G. Burchell, C. Gordon and P. Miller, eds. London: Harvester Wheatsheaf, pp. 87–104.

Freud, S. 2003. *The Uncanny.* D. Mclintock, trans. London: Penguin.

Giblett, R. 1996. *Postmodern Wetlands: Culture, History, Ecology.* Edinburgh: Edinburgh University Press.

Giblett, R. 2008a. *The Body of Nature and Culture.* Basingstoke: Palgrave Macmillan.

Giblett, R. 2008b. *Sublime Communication Technologies.* Basingstoke: Palgrave Macmillan.

Giblett, R. 2011. *People and Places of Nature and Culture.* Bristol: Intellect Books.

Gray, C. 1989. The Cyborg Soldier: The US Military and the Post-Modern Warrior. In: *Cyborg Worlds: The Military Information Society.* L. Levidow and K. Robins, eds. London: Free Association, pp. 43–71.

Grenville, B. 2001. *The Uncanny: Experiments in Cyborg Culture.* Vancouver: Arsenal Pulp Press.

Haraway, D. 1985. A Manifesto for Cyborgs: Science, Technology, and Socialist Feminism in the 1980s. *Socialist Review*, 80, pp. 65–107.

Haraway, D. 1995. Cyborgs and Symbionts: Living Together in the New World Order. In: *The Cyborg Handbook.* C. Gray, ed. London: Routledge, pp. xi–xx.

Haraway, D. 2000. The Virtual Speculum in the New World Order. In: *The Gendered Cyborg: A Reader.* G. Kirkup, L. James, K. Woodward and F. Hovenden, eds. London: Routledge, pp. 211–245.

Hutchins, J. 1999. Warren Weaver Memorandum: 50th Anniversary of Machine Translation. *MT News International, Issue 22*, 8(1), pp. 5–6.

Keller, E. 1994. The Body of a New Machine: Situating the Organism Between Telegraphs and Computers. *Perspectives on Science*, 2(3), pp. 302–323.

Landes, D. 1969. *The Unbound Prometheus: Technological Change and Industrial Development in Western Europe from 1750 to the Present.* Cambridge: Cambridge University Press.

Levidow, L. and K. Robins, eds. 1989a. Introduction. In: *Cyborg Worlds: The Military Information Society.* L. Levidow and K. Robins, eds. London: Free Association, pp. 7–11.

Levidow, L. and K. Robins. 1989b. Toward a Military Information Society? In: *Cyborg Worlds: The Military Information Society*. L. Levidow and K. Robins, eds. London: Free Association, pp. 159–177.

L'Isle-Adam, V. de. 1981. *The Eve of the Future Eden*. M. Rose, trans. Lawrence, KS: Coronado.

Lykke, N. 2000. Between Monsters, Goddesses and Cyborgs: Feminist Confrontations With Science. In: *The Gendered Cyborg: A Reader*. G. Kirkup, L. James, K. Woodward and F. Hovenden, eds. London: Routledge, pp. 74–87.

Margulis, L. 1971. Symbiosis and Evolution. *Scientific American*, 225, pp. 49–57.

Margulis, L. 1981. *Symbiosis in Cell Evolution: Life and Its Environment on the Early Earth*. San Francisco: W. H. Freeman.

Margulis, L. 1998. *Symbiotic Planet: A New Look at Evolution*. New York: Basic Books.

Metz, C. 1982. *Psychoanalysis and Cinema: The Imaginary Signifier*. C. Britton, A. Williams, B. Brewster and A. Guzzetti, trans. London: Macmillan.

Neale, S. 1985. *Cinema and Technology: Image, Sound, Colour*. London: Macmillan.

Robins, K. and L. Levidow. 1995a. Socializing the Cyborg Self: The Gulf War and Beyond. In: *The Cyborg Handbook*. C. Gray, ed. New York: Routledge, pp. 119–125.

Robins, K. and L. Levidow. 1995b. Soldier, Cyborg, Citizen. In: *Resisting the Virtual Life: The Culture and Politics of Information*. J. Brock and I. Boal, eds. San Francisco: City Lights, pp. 105–113.

Tomas, D. 1995. Feedback and Cybernetics: Reimaging the Body in the Age of the Cyborg. In: *Cyberspace/Cyberbodies/Cyberpunk: Cultures of Technological Embodiment*. M. Featherstone and R. Burrows, eds. London: Sage, pp. 22–43.

Vertov, D. 1984. *Kino-Eye: Writings*. K. O'Brien, trans. A. Michelsen, ed. Berkeley: University of California Press.

Virilio, P. 1989. *War and Cinema: The Logistics of Perception*. P. Camiller, trans. London: Verso.

Virilio, P. 1994. *The Vision Machine*. J. Rose, trans. Bloomington: Indiana University Press.

Virilio, P. 2000. *A Landscape of Events*. J. Rose, trans. Cambridge, MA: The MIT Press.

Warming, E. 1909. *Oecology of Plants: An Introduction to the Study of Plant Communities*. Oxford: Clarendon Press.

Wiener, N. 1989. *The Human Use of Human Beings: Cybernetics and Society*. London: Free Association. First published 1954.

7 The uncanny and the fictional

The uncanny and the modern adult literary fairy tale

E. T. A. Hoffmann for Freud (2003, pp. 135 and 141) was the 'one writer who was more successful than any other in creating uncanny effects' and 'the unrivalled master of the uncanny in literature.' Freud uses Hoffmann's 'The Sand-Man' with its doll Olimpia to illustrate and substantiate this claim, though Freud reduces the story to a proof text for the castration complex. Similarly, in a radio talk entitled 'Demonic Berlin' presented in 1930, Walter Benjamin (2014, pp. 24, 26) invoked Hoffman's penchant for 'the bizarre, the unconventional, the eerie, the inexplicable' and for 'the eerie, spooky, uncanny.' Hoffmann in 'My Cousin's Window' for Benjamin (1999, p. 425) created 'the type of the *flâneur.*' The *flâneur*, like the detective, is a connoisseur of the uncanny, of the fascinating and horrifying, of the secret and hidden coming out into the open.

One story of Hoffmann's that neither Benjamin nor Freud consider is 'The Mines of Falun' first published in 1819, one century to the precise year before Freud's essay on the uncanny. The story centres on 'the melancholy Elis' (Hoffmann, 1991, p. 305). His mother's death had 'lacerated his heart' (p. 306). Elis is seduced from a profitable seafaring life of trade to labouring in the mines of Falun. He initially resists the overtures of an old miner inviting him to become a miner. It would be unimaginable for him to exchange 'the beautiful free earth, the cheerful sunny sky' to 'go down into the fearful depths of hell' (p. 308). The old man challenges Elis that those who trade on 'the surface of the earth' are not nobler than the miner whose 'skill and unflagging labour unlock nature's secret treasures' (p. 308). Hoffmann sets up early on the spatial and moral metaphysics of the environment for his story by contrasting heaven and hell, surface and depth, open and locked, known and secret, light and darkness, profiteering and labour, ignoble and noble, trade and treasure.

The same night Elis has a dream about 'darkly gleaming minerals.' The deeper he looked, the more 'he saw in the depths innumerable charming female forms' (p. 309). These 'maidens' are presided over by 'the majestic Queen' (pp. 309 and 319) of the underworld who is both a femme fatale and a projection of Elis' dead mother. Eventually after three days of being pursued by 'the strange figments of his dreams,' Elis is seduced by his eroticized vision of the female body of the depths of the earth and sets off for the mines of Falun (p. 310). His initial encounter with the mines is not promising for 'when he stood before the huge jaw of hell, his blood froze in his veins and he became numb at the sight of the fearful, blighted desolation' (p. 311). Hoffmann uses the tried and true tropes of the orally sadistic mine, the industrial wasteland and freezing horror.

The sight of the mine gets worse as:

> not a tree, not a blade of grass, was living in the barren, crumbled, rocky abyss. [. . .] In the abyss there were stones – slag, or burned-out ores – lying around in a wild jumble, and sulfurous gases rose steadily from the depths as if a hellish brew were boiling, the vapors of which were poisoning all of nature's green delights. One would believe that Dante had descended from here and seen the Inferno with all its wretched misery and horror.
>
> (p. 311)

The depths of the earth do not yet lead to heavenly treasure, but to hellish trash in which the qualities of solid, liquid, gaseous and heat are mixed up and not in their proper place.

The chief official of the district and a shareholder in the mine later tells Elis that there is an ancient belief that 'the mighty elements, among which the miner boldly reigns will annihilate him unless he exerts his whole self in maintaining his mastery over them' in his work 'in the earth and the fire' (p. 315). Mining is a salutary instance of the perils of the drive for mastery as even in mastering the elements of earth and fire, the elements of air and water are polluted and ruined, the earth eventually too, and the purity of fire sullied in sulfurous fumes. The qualities of the solid, liquid, gaseous and heat and the four elements of earth, water, air and fire are mixed up in and by mining. Mining's mastery of the elements is a temporary control for the extraction of wealth resulting in polluted earth, air and water and a monstrous abyss. Mastery leads to monstrosity.

The abyss is the (un)home of the monstrous. This is the case on land and at sea. When Elis 'looked down into the monstrous abyss' (p. 311) of the mine he remembers the story an old helmsman told him about a dream he had about 'the immeasurable abyss that yawned beneath him [in the ocean] so that he could see the frightful monsters of the depths in horrible embraces' (p. 312). By contrast, in Elis' dream of the depths of the earth the 'charming female forms [are] locked in [presumably beautiful] embrace' (p. 309). After recalling the helmsman's dream, the monstrous forms of the mine for Elis seem like the 'revolting monsters' of the ocean and he 'felt himself trembling with horror' (p. 313). In fact, the monsters of the mine are even worse than the monsters of the ocean as he soliloquizes, 'What are all the horrors of the ocean compared to the frightfulness that dwells in that barren rocky abyss!' (p. 313). At least the ocean is fertile whereas the mine is infertile. The mine is not just hell but 'stygian hell,' dark hell with the River Styx of the underworld.

Elis decides to leave Falun but is seduced into staying by the sight of Ulla, the beautiful daughter of the chief official and shareholder in the mine. The thought of her 'form hovering like a shining angel above him' makes him forget 'all the horrors of the abyss' (p. 317). Elis goes to work in the mine to become wealthy enough to marry Ulla. One day when he is 'working in the deepest bore, wrapped in [. . .] sulfurous fumes' (p. 317) he hears a knocking sound that seemed to be coming from deeper in the mine, but where no one was working so the sound 'seemed quite uncanny' (p. 317). The sound of the knocking and the odour of the fumes are vehicles and vectors of the uncanny. The senses of hearing and smell, rather than sight, are the senses for perceiving the uncanny. The secret source of the knocking comes into the open when Elis 'saw a black shadow beside him, and as a cutting blast of air scattered the sulfur fumes.' He recognizes the old miner, who taunts him with the prospect of a rich mineral lode and prophesizes that he will never marry Ulla.

Elis returns to the surface and tells the foreman about both his recent and previous encounters with 'the uncanny miner' (p. 318). The miner is uncanny in Freudian terms as he is fascinating and horrifying and in Schellingian terms he is uncanny too as he has secrets, or is a secret, that emerge in the story. The foreman realizes that 'what we relate about him here is more than a legend.' In other words, it is a myth that taps into the mythology about mining and the earth. Over a century before, this miner had worked in the mine. He was 'melancholy' and a successful miner to whom 'the richest lodes were revealed to him as if he possessed a special higher power.' As a result, 'a story arose that he was in league with secret powers who reign in the bowels of the earth and fuse metals' (p. 318). He 'prophesied that a disaster would occur if it was not true love for marvelous rocks and metals that impelled the miner to work.' Mining should come out of love for 'marvelous rocks and metals,' and not out of lust for them. Mining should elicit gratitude for the generosity of the earth and not be driven by greed and gluttony. The miner's warnings were not heeded and 'out of greed, the mines were constantly enlarged until finally [. . .] a frightful cave-in occurred' and the miner was killed. The ghost of the dead miner continued to haunt and taunt the living miners, including Elis who becomes consumed by greed.

After Elis is apparently not preferred as Ulla's suitor, he dashes off to the mine. He sees 'the desolate rocks' as 'fearful monsters, the frightful offspring of hell' (p. 319). He descends into the mine and he thinks he finds the rich lode which the old miner kept talking about. Elis' 'fateful dream' returns of 'the fields of paradise filled with marvelous metal flowers and plants on which gems flashing fire were hanging like fruit, blossoms and flowers' (p. 319). Paradise for Elis in his dream is not a living garden, but a mineral and metallic fantasy of a dead garden. The femme fatale Queen seizes him and he surrenders himself to 'the subterranean kingdom of precious stones and minerals' (p. 321). He becomes 'split in half' with 'his better, his true being [. . .] climbing down into the center of the earth and [. . .] resting in the Queen's arms, while he was seeking his dreary bed in Falun' (p. 321), as yet unoccupied by the as yet unmarried good woman

Ulla. Dull, everyday life cannot match his fantasy world in and of the mine, nor can the good woman compare and compete with the femme fatale. He has become delusional, fixated on the femme fatale Queen of the underworld.

When he is about to be married to Ulla, he goes off to the mine to mine the fantasized rich lode, but he dies in a cave-in, an uncanny instance of history repeating itself and of the return of the repressed. Fifty years later his apparently petrified body is discovered (p. 323 and 324). He literally became rock when he was buried and crushed to death by the cave-in. When the elderly Ulla presses him to 'her withered breast,' the apparently petrified youth begins to 'turn to dust' (p. 324) in fulfilment of the biblical prophetic curse that God utters to Adam and Eve when He expels them from paradise, 'You are dust and to dust you will return' (*Genesis* 3: 19).

Hoffmann's tale is not only a theological parable about the perils of human hubris and avarice, not only a romanticist fable of a femme fatale in the guise of the Queen of the underworld spilt with the good woman in the character of Ulla, but also an environmental meditation on mining and the earth that theologizes the dark underworld of the mine and moralizes about mining. The mine is figured as an orally sadistic monster who consumes miners in keeping with a standard, prevalent trope of the nineteenth century, such as in Emile Zola's *Germinal* (see Giblett, 2011, chapter 9). The greedy miner is also figured as a monster who greedily lusts after and consumes the good things of the earth, also to be found elsewhere in writing about mining (see also Giblett, 2011, chapter 9). In a world and at time mining and polluting the earth to death as the powerhouse of its capitalist economy, Hoffman's tale might be a timely warning about the dangers of lusting greedily after minerals and metals and a salutary invitation to love the earth. Hoffmann's tale might remythify mining and the earth in gratitude for its generosity.

The uncanny and the gothic vampire romance

Bram Stoker's *Dracula* is not only *the* Gothic vampire romance par excellence, but also an uncanny story and a story of the uncanny. The Gothic and the uncanny are intimately intertwined. Although the word 'uncanny' only appears three times during the tale (pp. 16, 93 and 345), it is steeped in the uncanny. Similarly, the word '*unheimlich*' also appears three times in Beal's reading of *Dracula* in his study of religion and its monsters (2002, pp. 124, 127 and 134) in which he touches on several aspects of the uncanny in it and discusses the uncanny more generally (pp. 4–5 and 7–8). The settings for the tale and Dracula himself are uncanny. Dracula's castle is gloomily Gothic and his English estate is revoltingly bad smelling. He is not human. Dracula is a draconic monster who preys mainly on women.

The tale and its eponymous anti-hero take place and are situated in the monstrous between the human and the divine, in the liminal zone between the living and the dead, on the threshold between east and west, in the twilight between night and day, in the mysterious zone between the known and the unknown, in the realm between modernity and the pre-modern, and crosses the frontier from the tamed into the wild. The tale enacts a return of the repressed of the latter to the former and a return to the repressed of the former to the latter.

On the first page Harker narrates how, in travelling from Munich to Budapest, he was 'leaving the West and entering the East' (Stoker, 2011, p. 5). When he arrives in 'the midst of the Carpathians Mountains' he describes it as 'one of the wildest and least known portions of Europe' (p. 5). And it still is. He approaches his destination 'on the dark side of twilight [. . .] practically on the frontier' (p. 7). And it still is. When he arrives at Dracula's castle the welcoming hand shake felt 'more like the hand of a dead than a living man' (p. 18). He is more like a statue than a living body as 'his cast of face made his smile look malignant and saturnine' (p. 26). He is an 'Un-Dead phantom' and 'King Vampire' (p. 344).

Dracula looks gloomy and so is the estate Dracula has bought in England. Harker describes how 'there are many trees on it, which make it in places gloomy, and there is a deep, dark looking pond or small lake' (p. 25). Luckhurst (p. xiv) notes how generically Wilkie Collins' *The Woman in White* was 'Stoker's obvious model for *Dracula*' with its use of letters, diaries and news reports. It was also his model for some of the settings with Collins' Blackwater Lake the model for Stoker's description of the pond or small lake at Carfax. In Dracula's castle and thrall Harker finds himself increasingly entangled and trapped in 'the unnatural, horrible net of gloom and mystery which seemed [to be] closing around me' (p. 34). Later he wonders 'how can I escape this dreadful thrall of night and gloom and fear?' (p. 46).

Harker's responses to the place occur in the register of the uncanny. He 'felt a little strange, and not a little frightened' (p. 14) on his journey to Dracula's castle. On his second day at the castle Harker writes that 'there is something so strange about this place and all in it that I cannot but feel uneasy' (p. 27). Harker later writes variations on this theme as 'this accursed place,' 'this horrible place' (p. 35), 'this hateful place' (p. 37), 'this dreadful place' and 'this cursed land' (p. 52). Mina Harker writing about Dracula's castle notes that 'there was something wild and uncanny about the place' (p. 345). By travelling to Transylvania Harker had travelled not only from the tamed into the wild, but also back in time, to a place and time before modernity that it could not modernize. This place is buried in the past. Harker writes how 'the old centuries had, and have [,] powers of their own which mere "modernity" cannot kill' (p. 37). In Freud's terms, Harker travels back to a time 'long ago' and a place 'long forgotten.'

On his first day at the castle Dracula tells Harker, 'Transylvania is not England' (p. 23). Indeed, Transylvania is the antithesis of England: pre-modern to modern; darkened to enlightened; backward to progressive; slow to fast; and rural backwater to uncanny city. Dracula is the bridge between the two worlds, or at least between pre-modern Transylvania and modern London as he tells Harker that through reading books 'I have come to know your great England; and to know her is to love her. I long to go through the crowded streets of your mighty London, to be in the midst of the whirl and rush of humanity, to share its life, its change, and all that makes it what it is' (p. 22). Dracula feminizes England and longs to suck the life out of its modern, uncanny capital as if it were another vampyric victim. Dracula wants to be a *flâneur* in and of the crowd.

Lurking in the pre-modern twilight between the Gothic and the uncanny is the monster and the monstrous. In the case of *Dracula*, the monster is the Count himself. Even the name 'Dracula' signifies the monstrous. In his study of religion and its monsters, Beal (2002, p. 83; see also p. 125) traces how Dracula's 'name comes from the Romanian *dracul*, "the dragon," or "devil."' The dragon and the devil have haunted the European cultural imaginary for a long time. They are found in the Bible and in much literature thereafter, such as the Old English *Beowulf*, the legend of St George and the Dragon, Bunyan's *Pilgrim's Progress*, C. S. Lewis' *The Pilgrim's Regress* and J. R. R. Tolkien's *The Hobbit* (see Giblett, 2018, chapters 3 and 6). These are all Christian allegories and so is *Dracula* with Luckhurst (in Stoker, 2011, p. xix) even going so far as to suggest that *Dracula's* 'Christian allegory overtly replays St George and the slaying of the *drakul*, or dragon.' When Harker, the central narrator, is about to leave London for Dracula's Castle on the eve of St George's day, his wife warns him not to go telling him, 'Do you not know that to-night, when the clock strikes midnight all the evil things in the world will have full sway?' (Stoker, 2011, p. 8). In the tried and true conventions of horror fantasies, they do.

Dracula is a dragon not only by name, but also by nature. He leaves his castle at night on his vampyric missions not by leaving through a door, but by climbing out a window. Harker describes how Dracula 'crawl[ed] down the castle wall over that dreadful abyss *face down*' in what Harker calls 'lizard fashion' (p. 35; emphasis in original). Dracula is not human as he does not climb down backwards facing up, but crawls down facing forward. He upsets the order of things by inverting the upper and lower bodily strata. His face in the place where his grotesque lower bodily stratum, or nether regions, should be, and vice versa.

Like a dragon, he has wings 'with his cloak spreading around him like great wings' (p. 35). Unlike a dragon, he does not breathe out fire, but sucks in blood. He tells Harker on his first night at the castle, 'I have dined already. And I do not sup' (p. 19). He could have added, 'But I do suck,' but that would have given the game away too early. Later, after a night of vampyric feasting, Harker describes how 'the monster,' as 'if the whole awful creature were simply gorged with blood [. . .], lay like a filthy leech' (p. 51). Gorge means throat. Like the dragon whose fire comes up in his throat and out of his mouth consuming his victims, Dracula is orally sadistic.

By consuming blood Dracula is demonic because he violates several prohibitions against it in the Old Testament and arrogates to himself a privilege reserved for God. Beal (2002, p. 219, n.24) notes that 'some passages [in *Dracula*] indicate that God alone consumes the blood of sacrificed animals.' Beal cites several 'biblical prohibitions against blood-eating' in this endnote. *Dracula* itself cites 'the scriptural phrase, "for the blood is the life"' (Stoker, 2011, p. 218). This is the second part of *Deuteronomy* 12: 23. The first part of this verse states 'only be sure that thou eat not the blood' (cited by Beal, 2002, p. 130). Beal (2002, p. 130) comments that 'blood is life, flesh is not. While people are allowed to eat flesh, all blood, that is life, is God's and God's alone.' God created life, so He has the exclusive right to consume it. Dracula would become as God by eating blood, but in the process becomes the devil.

Dracula can also transform himself into 'nothing but a faint vapour' that 'trails under the door' (p. 262) and escapes when Harker's associates think that they have him trapped. He is evanescent and virtual like the crocodiles in L. G. Moberly's story that Freud thought was extraordinarily uncanny that transformed from solid, carved figures in wood into vague forms sliding on their bellies or 'underneaths.' Dracula does not have much solidity in the first place and transforms himself from vague form into faint vapour.

Moberly's story also highlights the importance of the sense of smell as a vehicle and vector for the uncanny. *Dracula* also has plenty of bad smells and 'malodourous air.' When Harker enters Carfax he writes:

> The long disuse had made the air stagnant and foul. There was an earthy smell, as of some dry miasma, which came through the fouler air. But as to the odour itself. How shall I describe it? It was not alone that it was composed of all the ills of mortality and with the pungent, acrid smell of blood, but it seemed as though corruption had itself become corrupt. Faugh! It sickens me to think of it. Every breath exhaled by that monster seemed to have clung to the place and intensified its loathsomeness.
>
> (p. 233)

Dry miasma are the bad or polluting airs rising from polluted earth. Miasma were thought to cause malaria (literally 'bad air') in the miasmatic theory of disease that prevailed in medicine from Hippocrates to the late nineteenth century in the decade when Stoker was writing *Dracula*, though he is more concerned with mental illness, metaphysical disease and the miasmatic theory of melancholy and madness than physical disease (see Giblett, 1996, chapters 5 and 7).

The uncanny and the detective story

The detective story reeks of the uncanny. The crime, especially if it is a murder, that precipitates the story is often uncanny. For Theodor Reik (cited by Vidler, 1992, p. ix), 'The unsolved murder is uncanny.' The quest to solve the mystery of who committed the murder is also uncanny, as is the setting where the murder took place and where the solution is to be found. The uncanny is the modus operandi of the private detective who has to descend into a place of almost impenetrable evil and darkness which can only be pierced by the penetrating light of reason brought to bear upon it by the superior intellect and insight of the great man himself. This place can either be a place made with human hands, such as the city, usually its dark underside, or a place not made with human hands, or made with ancient human hands, such as the swamp or other wetland figured as the slough of despond. This place is theologized as a place of evil, the monstrous and the demonic. The ostensibly most secular of modern literary forms in the detective story is short through with theology and needs to be detheologized and remythified.

In *The Hound of the Baskervilles*, probably Arthur Conan Doyle's most famous Sherlock Holmes story, the most famous private detective of them all solves the murders occurring in the uncanny Great Grimpen Mire. In Fergus Hume's even better-selling detective novel, *The Mystery of a Hansom Cab*, set in part in the slums of nineteenth-century Melbourne, the detective, like the epic hero and the Christian pilgrim, enters the infernal regions displaced into the slums of the city figured as the biblical 'valley of the shadow of death,' an uncanny place. This 'uncanny valley,' a concept developed by Masahiro Mori in 1970 as 'the trough we fall into when we die,' is inhabited by monstrous creatures who live in the liminal zone between the human and non-human, life and death.

Great Grimpen Mire

When Stapleton takes Dr Watson on a tour of the moor, he asks Watson if he notices '"those bright green spots scattered thickly over it."' Watson replies, '"Yes, they seem more fertile than the rest,"' to which his guide laughs and goes on to say, '"That is the great Grimpen Mire"' which he describes as '"the bog-hole"' and as '"an awful place."' By acknowledging the fertility of the mire, Watson was a wetlands ecologist before his time! Whilst the mire is being discussed:

> A long, low moan, indescribably sad swept over the moor. It filled the whole air, and yet it was impossible to say whence it came. From a dull murmur it swelled into a deep roar, and then sank back into a melancholy, throbbing murmur once again.

Watson asks what the sound is only to be told that '"the peasants say it is the Hound of the Baskervilles calling for its prey."'

Watson is not satisfied with this answer, dismissing it as '"nonsense"' and pressing Stapleton for a more rational explanation which he feebly offers:

> 'Bogs make queer noises sometimes. It's the mud settling, or the water rising, or something.' 'No, no that was a living voice.' 'Well, perhaps it was. Did you ever hear a bittern booming?' 'No, I never did.' 'It's a very rare bird – practically extinct – in England now, but all things are possible upon the moor . . .' 'It's the weirdest, strangest thing that ever I heard in my life.' 'Yes, it's rather an uncanny place altogether.'

Indeed, the wetland is the uncanny place *par excellence* for Western culture (as argued in a previous chapter of this volume).

Watson is greatly affected by 'the melancholy of the moor' and reports to Holmes from what he calls:

> this most God-forsaken corner of the world. The longer one stays here the more does the spirit of the moor sink into one's soul, its vastness, and also

> its grim charm. When once you are out upon its bosom you have left all traces of modern England behind you.

This conclusion is hardly surprising because once you have entered the mire mother, or the mother of all mires, you have entered a pre-modern, even pre-Christian, landscape, or more precisely wetlandscape, which is the precursor of the modern landscape and the 'foundation' for it. Even though one may avoid sinking into it physically, it sinks into one metaphysically and psychologically. It is not a settled and stable object, but its own free agent, a dynamic process which cannot be kept within strict bounds, whose primary horror is that it breaks boundaries.

In the conclusion to the story Holmes and Watson follow Stapleton into 'the widespread bog' with its 'green-scummed pits and foul quagmires' where:

> rank weeds and lush, slimy water-plants sent an odour of decay and a heavy miasmatic odour onto our faces, while a false step plunged us more than once thigh-deep into the dark, quivering mire. [. . .] Its tenacious grip plucked at our heels as we walked, and when we sank into it, it was if some malignant hand was tugging us down into those obscene depths, so grim and purposeful was the clutch in which it held us.

Although Watson as a modern medical man does not seem to subscribe to a miasmatic theory of malaria, he supports a miasmatic theory of melancholia. He also anthropomorphizes the mire by giving it hands to pull Holmes and him down into its clutching, (s)mothering embrace as it had pulled Stapleton 'down in the foul slime of the huge morass.'

The valley of the shadow of death

Similarly, Fergus Hume in his best-selling detective novel and 'the first popular Melbourne novel,' according to John Arnold (1983, pp. 7–8), *The Mystery of a Hansom Cab*, first published in 1886, evokes vividly the Dantesque 'Infernal Regions' 'off Little Bourke St.' Like the epic hero descending into the underworld to slay monsters and return home triumphant, 'the detective led the way down a dark lane, which felt like a furnace owing to the heat of the night' (Hume, 1999, pp. 178–179). The infernal distress of the slums of Bourke Street can be contrasted with what Marjorie Clark (in Arnold, 1983, p. 69) in 1927 described as 'the cool charm of its [Collins Street's] arcades.' Worse for the detective entering the slums of Bourke Street, 'it was like walking in the valley of the shadow of death. [. . .] And, indeed, it was not unlike the description in Bunyan's famous allegory what with the semidarkness, the wild lights and shadows' (Hume, 1999, p. 180). Hume combines Dante and Bunyan (and the biblical Psalmist, as Bunyan is drawing on *Psalm* 23) in an omnibus and ominous grab bag of tropes for the urban underworld.

As an aside here, these pages are excerpted and reproduced as a quintessential depiction of the 'mean streets and back alleys' of late nineteenth-century

Melbourne as depicted in crime fiction for the anthology *Literary Melbourne* edited by Stephen Grimwade (2009, pp. 187–190). The reference to 'the valley of the shadow of death' is later misquoted as 'the shadow of the valley of death' (Grimwade, 2009, p. 244). This misquotation gives a new nuance to the idea of death valley transposed here from the remote deserts of California to inner-city Melbourne. Both are hot and deadly places. As Michael Cannon (1976, pp. 42 and 43) puts it, 'The Angel of Death came early and stayed late in the Melbourne of 1892 and 1893' with 'epidemics of influenza, typhoid and measles [. . .] which [. . .] killed thousands.' It is only a hop, skip and a jump from Bunyan's and the Psalmist's 'valley of the shadow of death' to Hume's alley of the shadow of death and Grimwade's shadow of the valley of death.

The detective's journey into the urban underworld is not only physical but also moral and allegorical. The detective novel is a secular allegory of literal and spiritual descent and ascent, of degradation and salvation, just like the sacred allegory of Bunyan's *Pilgrims Progress.* In the case of Sherlock Holmes, the detective's journey also includes a descent into the Bunyanesque 'Slough of Despond' of great Grimpen Mire in *The Hound of the Baskervilles* (as we have already seen). Similarly, Flinders Street in 1845 was described as 'that slough of despond' (cited by Annear, 2014, p. 40). The swamp outside the city in the country, or inside the city in the muddy streets or slums, is a secularized satanic space. Perhaps it is fitting that the *flâneur* is a creature of the arcades, which Johann Geist (1983, p. vii) describes in his monumental history of the arcade as 'a secularised sacred space.' The swamp is a secularized satanic space for the detective that is a sacred space and place for indigenous, traditional peoples. The secularized satanic space of the swamp could be the native or natural swamp in the country, or the feral or cultural 'swamp' of the slums in the city in which the former is used as a figure for the latter.

In *Pilgrim's Progress* Bunyan (2008, p. 65) takes the fourth verse of *Psalm* 23 referring to the valley of the shadow of death for his text and sermonizes on how 'Christian must needs go through it, because the way to the Celestial City lay through the midst of it.' Typically the detective in the modern city must go through the valley of the shadow of death of the underworld of the slums in order to protect the upper city of the upper ten thousand living in the upper world of the crystalline celestial city. Or the hero of the modern epic of the novel, such as Jean Valjean in Hugo's *Les Miserables,* must descend into the underworld of the sewers of Paris (see Giblett, 2016, chapter 3). The modern hero of the tourist can even follow in Valjean's footsteps and go on a tour of the sewers of Paris in the marshy underworld of 'Lutetia,' the Latin-cum-Celtic name for the 'filthy marsh' in which the city was founded. Although Melbourne as the 'Paris of the South' does not have such an official tour of its sewers, 'a small group has for several decades mounted regular unlawful explorations of Melbourne's stormwater drains and tunnels' in what Kristin Otto (2005, p. 175) calls 'a Yarra [River] underworld.' Sophie Cunningham (2011, p. 147) also comments that 'while Melbourne does not boast Paris' hundreds of kilometres of underground tunnels, it does have Anzac, a cavernous drain under South Yarra where many parties have been held over the years.'

Later in *Pilgrim's Progress* Christian is informed that 'the valley [of the shadow of death] itself' is:

> as dark as pitch: we also saw there the hobgoblins, satyrs, and dragons of the pit; we heard also in that valley a continual howling and yelling, as of a people under unutterable misery, who there sat bound in affliction and irons; and over that hung the discouraging clouds of confusion; Death also does always spread his wings over it. In a word, it is every whit dreadful, being utterly without order.
>
> (Bunyan, 2008, pp. 67 and 69)

As the underworld is a monstrous place itself inhabited by monsters, such as dragons, so is 'the monster city.' In similar vein, Little Bourke Street in *The Mystery of a Hansom Cab* has its 'weird and grotesquely horrible' inhabitants (Hume, 1999, p. 220). In a word, they are monstrous. Little Bourke Street contrasts with Bourke Street proper not only in the type of inhabitants, but also in the mode of illumination. Bourke Street is described 'as the brilliantly lit street' produced by 'electric lights,' which highlight the members of the crowd caught in 'the full glare of the electric light.' By contrast, Little Bourke Street is lit by 'sparsely scattered gas lamps with their 'dim light,' whereas the lanes have no lights or lamps so they are dark, or:

> not quite dark, for the atmosphere had that luminous kind of haze so observable in Australian twilights, and the weird light was just sufficient to make the darkness visible.
>
> (Hume, 1999, pp. 178–180)

Typically the fictive detective of the modern city enlightens the benighted, brings light to darkness and illuminates the crepuscular gloom of crime and grime.

In her biographical and critical study of Hume's novel, Lucy Sussex (2015, p. 7) argues that it had 'an important role in establishing detective fiction as a publishing category' as he is 'one of the most influential crime writers of all time' who wrote 'the biggest-selling crime novel of the nineteenth century, and one of the most important Australian books ever.' Sussex traces the etymology of the word 'detective' and how it literally means 'de-roofing' as 'a detective raises the roof, figuratively.' The detective in the detective story, and the detective storyteller, raises the roof of dwelling spaces, looks inside and reveals what is inside to the reader. The detective and detective story reader are positioned as snooping voyeurs.

Sussex (2015, pp. 9–10) also traces how this uncovering and de-roofing had a demonic function and cites Dickens in *Dombey and Son*, published in 1848, who pleaded for 'a good spirit who would take the house-tops off.' David Grann

(2017, p. 57) similarly traces how 'the term "to detect" derived from the Latin verb to "unroof," and because the devil, according to legend, allowed his henchmen to peer voyeuristically into houses by removing their roofs, detectives were known as "the devil's disciples."' In the Sherlock Holmes' story 'A Case of Identity,' Arthur Conan Doyle has Holmes describe the detective's work of unroofing in similar terms (albeit without the etymology and the theological overtones). The detective in the detective story, and the detective storyteller too, perform these socially useful, but morally ambiguous, roles of the good, or demonic, spirit. Both were *flâneurs* who entered 'the city's central hell [of] the slums,' as Sussex (2015, p. 60) puts it in relation to Hume and *Hansom Cab*. As with the circles of Dante's hell, or inferno, the city has a centre, and its centre is the lower depth of its slums and sewers, the grotesque lower urban strata. The seemingly most secularized literary genre of the detective story about the modern city has strong theological undertones.

The uncanny and the weird horror story

H. P. Lovecraft is the twentieth-century master of the uncanny in fiction to rival Hoffmann in the nineteenth century, not just because the word 'uncanny' appears regularly in his fiction (usually once per story, more in the longer pieces) and not just because he probably coined the adverb 'uncannily' in the 1920s (*The Shorter Oxford English Dictionary* is not forthcoming on this score), but because he engages with the monstrous and enters that liminal zone between it and the divine that evokes horror and fascination for his characters often via the sense of smell with foetid, fishy or rank odours, but interestingly not in 'The Moon-Bog.' Lovecraft's story uses the word 'uncanny' twice in this story (Lovecraft, 2004, pp. 46, 47 and 409). More than this superficial lexical choice, however, the story evokes horror and fascination in response to 'the thing' (p. 42) associated with, and residing in, an Irish bog, an uncanny place par excellence along with other wetlands.

Lovecraft was following in the boggy and marshy footsteps of one of his writer heroes, Lord Dunsany (1908, pp. 31–45), especially his story 'The Kith of the Elf-Folk' living in an Irish marsh. The Irish connection is also secured by the fact that 'The Moon-Bog' was written for a meeting of writers in Boston on St Patrick's Day in 1921. 'The Moon-Bog' relates the story of a wealthy American of Irish background who moves back to Ireland to restore the ruined castle located by the ancient bog on the family estate. 'The peasants' are initially delighted with the American's restoration of the castle and they 'blessed him for bringing back the old days with his gold from over the sea' (p. 42). He was bringing the past back into the present which does not auger well for any aficionado of the uncanny as one never knows what might be brought back from the past into the present.

The American also wanted to destroy the past and take the modern present back to the ancient past by draining the bog. Because of his foolhardiness, 'in time there came troubles, and the peasants ceased to bless him, and fled away instead as from a doom' (p. 42). 'Doom' is a Lovecraft code word for apocalyptic destruction and the end of time. 'Troubles' is a term that refers in part to the troubles between the Catholics and Protestants in Northern Ireland. The story can be read as a political allegory of the sectarian divide between the Catholics and Protestants in Northern Ireland, and between rural republican southern Ireland and industrial royalist Northern Ireland.

The American blames the bog as the cause of 'the peasants'' fleeing and giving him trouble, but they:

> had gone from Kilderry because Denys Barry [the American] was to drain the great bog. For all his love of Ireland, America had not left him untouched, and he hated the beautiful wasted space where peat might be cut and land opened up.
>
> (p. 43)

Kilderry is a nice touch as it alludes to both the real city of Kildare and the famous Irish creamy ale Kilkenny. How Irish can you get?! The American has been corrupted by the modern drive to drain wetlands and wishes to do so in an act of agricultural sadism and geographical surgery out of misaquaterrism, the hatred of wetlands.

Unlike his friend the first-person narrator, the ugly American who hates the bog does not have an aesthetic appreciation for 'the beautiful wasted space' of the bog. The narrator relates earlier how he:

> reached Kilderry in the summer sunset, as the gold of the sky lighted the green of the hills and groves and the blue of the bog, where on a far islet a strange olden ruin glistened spectrally. That sunset was very beautiful.
>
> (p. 42)

The narrator paints a pastorally idyllic and picturesque depiction of the pleasing prospect in which the bog is not a blot on the landscape, nor in the American's account books, but a vital, integral and beautiful part of it. The story can be read as an allegory of modern aesthetics that regards wetlands as ugly versus the traditional view of them as beautiful, or uncanny.

The story can also be read as an allegory of modern capitalist resource extraction against the traditional love of bogs tied up with 'legends and superstitions' (p. 43). When the American tells the first-person narrator about:

> the fears which had driven the people from Kilderry I laughed as loudly as my friend had laughed, for these fears were of the vaguest, wildest, and most absurd character. They had to do with some preposterous legend of the bog, and of a grim guardian spirit that dwelt in the strange olden ruin on the far islet I had seen in the sunset. There were tales of dancing lights in the dark of the moon, and of chill winds when the night was warm; of wraiths in white hovering over the waters, and of an imagined city of stone

> deep down below the swampy surface. But foremost among the weird fancies, and alone in its absolute unanimity, was that of the curse awaiting him who should dare to touch or drain the vast reddish morass. There were secrets, said the peasants, which must not be uncovered; secrets that had lain hidden [from past times long ago].
>
> (p. 43)

The ruined castle from the past could be restored to present glory, but 'the brooding bog' (p. 44) in and of the past should remain there and not be brought into the present by being drained.

Yet this is precisely what the American was setting out to do. After the peasants flee, the American brings in 'labourers from the north' to drain the bog:

> Now the work of drainage was ready to begin, and the labourers from the north were soon to strip the forbidden bog of its green moss and red heather, and kill the tiny shell-paved streamlets and quiet blue pools fringed with rushes.
>
> (p. 44)

The labourers from the north participate in the agricultural and voyeuristic sadism of stripping the bog of its clothing and in the American's aquaterracide (the killing of a wetland). The bog is personalized as a sentient being, if not feminized, the bride stripped bare by her haters, to paraphrase Marcel Duchamp.

The narrator is ambivalent about this act of environmental vandalism:

> For although I disliked to see the moss and the heather and the little streams and lakes depart, I had a growing wish to discern the ancient secrets the deep-matted peat might hide.
>
> (p. 45)

He is driven by the same epistemophilia for enlightenment that drives his host.

That night he awakes suddenly and sees out 'of a Gothic lattice window [. . .] a spectacle which no mortal, having seen it, could ever forget':

> The wide plain, the golden moonlight, the shadowy moving forms, and above all the shrill monotonous piping, produced an effect which almost paralysed me; yet I noted amidst my fear that half of these tireless, mechanical dancers were the labourers whom I had thought asleep, whilst the other half were strange airy beings in white, half indeterminate in nature, but suggesting pale wistful naiads from the haunted fountains of the bog.
>
> (p. 45)

The labourers have been turned into uncanny automatons like 'the bachelors' in Duchamp's 'The Bride Stripped Bare by Her Bachelors,' and the narrator is gripped by uncanny fascination and horror. Duchamp's work dates from

1915–1923, the same period as Freud's essay and Lovecraft's story. All three followed the heyday of bachelor machines from 1910–1914 (see Giblett, 2008a, pp. 95–98, 2008b, pp. 167–168).

Music rather than scent, the sense of hearing rather than the sense of smell, is the vector and vehicle of the uncanny in this story, as it is famously in Lovecraft's weird story 'The Music of Erich Zann' (Lovecraft, 2001, pp. 45–52 and 376–377), probably written in December 1921 after 'The Moon-Bog' in March of that year. In many other stories by Lovecraft the sense of smell with foetid, fishy or rank odours is the vehicle and vector for the uncanny.

The narrator's ambivalence about draining the bog turns to dread:

> For some unknown reason I dreaded the thought of disturbing the ancient bog and its sunless secrets, and pictured terrible sights lying black under the unmeasured depth of age-old peat. That these secrets should be brought to light seemed injudicious.
>
> (p. 46)

The enlightenment project of bringing the dark secrets of the past buried in the bog into the light of the modern present now seems ill-advised and highly dangerous. The American should have let sleeping bogs lie. The uncanny brings to present consciousness past 'legends and superstitions.' What come into the open in Lovecraft's story are the dangers of epistemophilia and enlightenment, and the evils of misaquaterrism and aquaterrism.

The following night the narrator has a similar vision, a recurring waking dream, or nightmare again seen through 'the Gothic window' (p. 47) for the whole story is placed within a Gothic frame and seen through a Gothic screen or lens:

> As I watched in awe and terror I thought I saw dark saltant forms silhouetted grotesquely against the vision of marble and effulgence. The effect was titanic – altogether unthinkable – and I might have stared indefinitely had not the sound of the piping seemed to grow stronger at my left. Trembling with a terror oddly mixed with ecstasy [. . .]
>
> (p. 47)

The narrator is experiencing the sublime of the lost city buried beneath the bog limned with the slime of the uncanny bog. Slime is the secret of the sublime as expressed in the portmanteau word s(ub)lime and as invented by Zoë Sofoulis (cited by Giblett, 1996, p. 27).

The narrator sees:

> Half gliding, half floating in the air, the white-clad bog-wraiths were slowly retreating toward the still waters and the island ruin in fantastic formations suggesting some ancient and solemn ceremonial dance. Their waving

> translucent arms, guided by the detestable piping of those unseen flutes, beckoned in uncanny rhythm to a throng of lurching labourers who followed dog-like with blind, brainless, floundering steps as if dragged by a clumsy but resistless daemon-will.
>
> (pp. 47–48)

The bog-wraiths seem to be gliding like flying creatures and floating like swimming creatures. They are half fowl and half fish, hybrids. They belong to neither the element of air, nor of water, but cross the liminal zone between them just as the uncanny automatons of the labourers cross the liminal zone between the human and non-human (dogs, daemons, machines).

The narrator:

> fled from that accursed castle along the bog's edge[.] I heard a new sound; common, yet unlike any I had heard before at Kilderry. The stagnant waters, lately quite devoid of animal life, now teemed with a horde of slimy enormous frogs which piped shrilly and incessantly in tones strangely out of keeping with their size. They glistened bloated and green in the moonbeams, and seemed to gaze up at the fount of light.
>
> (p. 49)

Unlike most other Lovecraft tales, the smell of the stagnant waters is not described either for good or ill. Typically, they would be described for ill as foetid odours, fishy or rank, but this story leaves the smell of the bog unremarked.

The narrator's final sight of the bog, however, is elaborated:

> I followed the gaze of one very fat and ugly frog, and saw the second of the things which drove my senses away. Stretching directly from the strange olden ruin on the far islet to the waning moon, my eyes seemed to trace a beam of faint quivering radiance having no reflection in the waters of the bog. And upward along that pallid path my fevered fancy pictured a thin shadow slowly writhing; a vague contorted shadow struggling as if drawn by unseen daemons. Crazed as I was, I saw in that awful shadow a monstrous resemblance – a nauseous, unbelievable caricature – a blasphemous effigy of him who had been Denys Barry.
>
> (p. 49)

The bog with its monstrous frogs has its revenge against the monstrous American who wanted to drain it. Rather than the bog being monstrous, or being inhabited by a monster, the modern American is monstrous. The narrator's opening gambit for the story is to entice the reader with the enigma of 'somewhere, to what remote and fearsome region I know not Denys Barry has gone' (p. 42). He has certainly gone into the bog, but presumably he has also gone into the regions far below the bog of the underworld never to return. He has received his just deserts for his misaquaterrism and aquaterracide in the triumph of good over evil and wisdom over folly, the moral of the story as in all good fairy tales: Long live bogs!

The uncanny and the dystopian science fiction novel

In J. G. Ballard's second novel, *The Drowned World*, first published in 1962, the city of London has been transformed into a tropical lagoon brought about by the increased activity of solar flares melting the ice caps. In the age of global warming and climate change *The Drowned World* now seems prescient. It has given birth to a whole series of dystopian science fiction novels about drowning cities, none of which have the same reach and depth as *The Drowned World*, especially when it comes to the uncanny and the return of and to the repressed of the waters below and before the city, including its wetlands (see Giblett, 2016). In Ballard's novel some of London's skyscrapers have been left with their tops sticking up out of the water. The world below has been drowned, but the skyscrapers continue to serve a vestigial function for which they were designed – to transcend the world below, to sublimate the base matter of the earth below (the slime) into the supreme heights of abstraction (the sublime) repressing what is below (see Giblett, 1996, Figure 1, chapter 2). *The Drowned World* is concerned with the return of the spatial repressed (the tropical colonies and the uterus), temporal repressed (the uterine and colonial pasts) and psychological repressed (memories of the womb and the mother's body). Indeed, for *The Drowned World* there is a homology between all three.

In *The Drowned World* the previously drained world, the temperate dry lands, of London is flooded as a return of the repressed and colonized tropical 'other,' the tropical wetlands. In *The Drowned World*, a return of the repressed without, the exocolonized lagoons of the tropics, occurs. The boundary between the tropical and the temperate ceases to exist in *The Drowned World* with the former's region and climate having taken over the latter's. London, and the whole of England for that matter, is now located in 'the European lagoons' (p. 12).

The inhabitants of London are now so many latter-day colonists of the tropical 'at home.' The lagoon which was not 'at home,' but was colonized in the tropical, has come home; the wetlands which were not like those 'at home' have come home and made home their own; the unhomely has invaded and conquered the homely. In this return of the colonial repressed the official job of Kerans, the central character of the novel, is to serve a vestigial colonial function of cartographer by 'mapping the shifting keys and harbours and evacuating the last inhabitants . . . living on in the sinking cities' (p. 12). As the story progresses, however, Kerans is transformed from imperialist cartographer to wetlands conservationist. By an etymological play on 'lagoon' and 'lacuna,' both of which come from the Latin '*lacuna*' meaning pool, Kerans' job is made more difficult because a lacuna (and a lagoon on a map) is 'a hiatus, blank, missing part, a gap, an empty space, spot, or cavity.' For the cartographer the lagoon, the slime (and the wetland more generally) is an impossible abject because of its changing shape.

Yet Kerans undergoes a transformation in his perception of the wetland. Initially, for him 'the lagoon was nothing more than a garbage-filled swamp' (p. 13). Culturally, and spatially, the two places are not so far apart as wetlands have often been the site of 'sanitary land fill.' The description of the swamp as 'nothing more . . .' upholds the Western tradition of reductive, pejorative markers being assigned to wetlands. Riggs concurs with Kerans' summation when he pronounces that 'the whole place is nothing but a confounded zoo' (p. 17), which is like Jefferies' reduction of the city to the animal. Ballard's lagoon smells bad: 'Kerans felt the terrible stench of the water-line, the sweet compacted smells of dead vegetation and rotting animal carcasses' (p. 13). Kerans feels, rather than smells, the stench. Perhaps he would rather feel the stench than smell it. The sensation of touch stays on the outside whereas the sensation of smell penetrates inside. For Kerans, 'A thick cloacal stench exuded from the silt flat' (p. 61).

The returns of the repressed, of the bad smells of the swamp, of the colonial other, also involve the return to the past, both the past of the human species and the individual past, in a reversal of the Freudian tenet following Haeckel that individual development repeats the development of the species (ontogenesis repeats phylogenesis). In *The Drowned World*, the apocalypse of the world drowning reverses phylogenesis in a return to the primeval slime, or more precisely to 'impenetrable Mato Grossos sometimes three hundred feet high . . . a nightmare world of competing organic forms returning rapidly to their Paleozoic past' (p. 19).

Ontogenesis is also reversed for Kerans in a return to amniotic fluid, a prospect which fills him initially with horror when he wonders whether 'perhaps these sunken lagoons simply remind me of the drowned world of my uterine childhood – if so, the best thing is to leave straight away' (p. 28). The lagoon, perhaps of all wetlands, is markedly, though naturally, enclosed. On a continuum of wetland types from the most earthbound to the least, the lagoon is the most wet of the wetlands. It is also separated or closed off from the sea by low sandbanks or by an atoll, a small sea within a larger. The lagoon is a kind of natural vessel made, like all wetlands, of earth and water, but with a thinner shell, a rim and even a lip. The drowned world is referred to as 'the black bowl of the lagoon' (p. 47, see also p. 70). No doubt the black bowl contains black water.

For Kerans the drowned worlds of the phylogenetic and ontogenetic, of the exterior and interior, past and present become indistinguishable. The uterine past of the individual and the paleozoic past of the human species have been repressed. Their joint return are represented in the iconography of the drowned world itself and portrayed in 'one of Max Ernst's self-devouring phantasmagoric jungles [which] screamed silently to itself, like the sump of some insane unconscious' (p. 29). The unconscious is even more strongly figured in aquatic terms in the reference to 'the submerged levels below his consciousness' (p. 83).

Rivers have been masculinized and swamps feminized, the latter most powerfully in the organizing principle of *The Drowned World*, the wetland as womb, though the novel develops this trope from the modality of horror to that of fascination. Rather than enacting a typically masculinist repudiation of his own birth, Kerans becomes fascinated with his uterine experience, though not with his mother per se. He shows no interest in his mother as a person. His fascination with his uterine experience is abstracted from the body in which it occurred. Kerans' mother is conspicuous by her absence from his memory and his curiosity. She is a hiatus, blank, missing part, a gap, an empty space, in short, a lacuna in the story of his life, a lagoon in the map of his life.

Perhaps not surprisingly then, his attitude to Beatrice, the only female character in the novel, is inflected with misogyny. The choice of name is rather ironic given Dante's fascination with another Beatrice and with the slimy in Canto 7 of the *Inferno* (see Giblett, 1996, chapter 2). The wetland hell of *The Drowned World* could be read as a post-modern version of Dante's medieval vision of a Stygian marsh with its souls of the sullen stuck in slime. For Kerans, though, the Beatrice of *The Drowned World* is not the beatific Beatrice of Dante's *Divine Comedy*, but 'Pandora with her killing mouth and witch's box of desires and frustrations, unpredictably opening and shutting the lid' (p. 31). Beatrice is figured as an automaton, that ultimate figure of the *unheimlich* or uncanny for Freud, with a devouring, mechanical and orally sadistic mouth.

Kerans is a hero who literally descends into the underworld of the drowned world, or womb, of London and not only survives but undergoes a transformation, or kind of rebirth. Kerans' fascination with his uterine past and its homologous relationship with the Paleozoic past of the human species are confirmed by Bodkin: 'Just as psychoanalysis reconstructs the original traumatic situation in order to release the repressed material, so we are now being plunged back into the archaeopsychic past, uncovering the ancient taboos and drives that have been dormant for epochs' (pp. 43–44). Plunged back, in other words, into the repressed world of primeval slime and the swamp as a kind of recapitulation of birth in reverse and in the process raising the unconscious to consciousness in a successful 'talking cure.'

Bodkin goes on to suggest that 'the uterine odyssey of the growing foetus recapitulates the entire evolutionary past' (p. 44), or in other words, ontogenesis recapitulates phylogenesis from primeval slime to post-modern times as a heroic journey. Individually, and ontogenetically, for Bodkin, 'we really remember these swamps and lagoons' (p. 48). The lacunae in memory and the lagoon on the map are filled up. The lagunine apocalypse of a decaying Europe, however, recapitulates in reverse the entire evolutionary past from post-modern times back to primeval slime to arrive at 'the very junction where we stand now on the shores of this lagoon, between the Paleozoic and Triassic Eras' (p. 44). This is both a form of time-travel in which the earth (and its history) is traced temporally (which includes its wetlands unlike the spatial tracing of the earth (and its geography) on maps which invariably do not trace wetlands) and of what Bodkin graces with the high-sounding name of 'the Psychology of Total Equivalents' or 'Neuronics':

> I am convinced that as we move back through geophysical time so we re-enter the amnionic [*sic*] corridor and move back through spinal and

> archaeopsychic time, recollecting in our unconscious minds the landscape of each epoch, each with a distinct geological terrain, its own unique flora and fauna, as recognisable to anyone else as they would be to a traveller in a Wellsian time-machine. Except that this is no scenic railway, but a total re-orientation of the personality. If we let these buried phantoms master us as they re-appear we'll be swept back helplessly in the flood-tide like pieces of flotsam.
>
> (pp. 44 and 45)

Like Kerans, Bodkin fears being plunged or swept back in metaphors of fluidity into the past, whether it be the uterine or paleozoic past or both. It is a matter of master or be mastered, or more precisely mistressed. Even though the tropical and the orient have won out climatically and geographically over the temperate and the occident, the battle still rages psychologically between these sites in the individual and collective mind and body. The colonies have struck back and triumphed externally, but internally the battle still goes on.

Much of that battle is fought out, and lost, as would be expected in accordance with Freud, in the domain of dreams. Kerans has a dream in which he steps out into the lake:

> whose waters now seemed an extension of his own bloodstream. As the dull pounding rose, he felt the barriers which divided his own cells from the surrounding medium dissolving, and he swam forwards, spreading outwards across the black thudding water.
>
> (p. 71)

This dream of the dissolution of Kerans' ego in black water is a reversal of the dreams which symbolize the ego's formation. In Kerans' dream he leaves the security of the castle of the self and flounders around in the black waters and garbage dump of London's lagoons. This dream is a recurring one which does not just involve the dissolution of the boundaries of his body and his ego, but also a return to his amniotic past. The lake becomes 'the warm amnionic [*sic*] jelly through which he swam in his dreams' (p. 99). Yet this womb associated with life is also ambivalently, and misogynistically, figured as a tomb associated with death.

All Kerans' fears of ingestion and dissolution are re-enacted when, in the ninth chapter entitled, 'The Pool of Thanatos' (death-instinct), he dives into 'a tank filled with warm, glutinous jelly that clamped itself to his calves and thighs like the foetid embrace of some gigantic protozoan monster' (p. 104). Given that he is diving into water figured as amniotic fluid, the body of his mother is, by implication, 'a protozoan monster' and his uterine childhood a foetid, (s)mothering embrace from which he has escaped and to which he does not want to return. When he is below he is asked, 'How's the grey sweet mother of us all?' to which he replies 'Feels like home' (p. 105), but an unhomely or *unheimlich* home which is uncanny.

Kerans descends further into this uncanny home by entering an auditorium 'with its blurred walls cloaked with silt [which] rose up above him like a huge velvet-upholstered womb in a surrealist nightmare' (p. 108). Yet Kerans almost dies from diving too deep, but in the process experiences a kind of rebirth and the realization of his own dream of dissolution of his ego and body boundaries:

> As the spotlight flared across the domed ceiling, illuminating the huge vacant womb for the last time, Kerans felt the warm blood-filled nausea of the chamber flood in upon him. He lay back [. . .] the soothing pressure of the water penetrating his suit so that the barriers between his own private bloodstream and that of the giant amnion seemed no longer to exist. The deep cradle of silt carried him gently like an immense placenta, infinitely softer than any bed he had ever known.
>
> (p. 110)

When Kerans' dream is actualized, it is not as horrific as he feared it might have been, and the embracing, entangling protozoan mother-monster turns out to be an immense, soft placenta with living black waters. Kerans' view of wetlands changes from oral-sadistic, consuming and death-dealing *vagina dentata* to corporeal-pleasurable, (re)producing and life-giving placenta. The silt is no longer horrific and nightmarish, but comforting and nurturing. The velvet-upholstered womb in a surrealist nightmare becomes a deep cradle in a pleasant daydream. The auditorium becomes transformed in recollection into 'the great womb-chamber of the planetarium' (p. 112), an exterior architectural analogue, or homologue, of interior bodily space.

So when Strangman, the villain of the story, undertakes the project of pumping dry a section of London, 'turning the once limpid beauty of the underwater city into a drained and festering sewer,' Kerans is 'unable to accept the logic of the rebirth before him' (p. 121), a rebirth unlike the one which he has just experienced in that instead of returning to the deep, to the amniotic, as he did when he dove into it, part of the lagoon is pumped dry in order to try to return London to its pre-drowned state. For Beatrice, the drained section of London is 'like some imaginary city of Hell' (p. 123), a necropolis, in fact (p. 151). This Hell is Dantesque, or at least like the fifth circle of Dante's *Inferno* set in the slime of the Stygian marsh, especially with 'the black slime oozing' out of the pores of the protozoan mother-monster:

> Everything was covered with a fine coating of silt, smothering whatever grace and character had once distinguished the streets, so that the entire city seemed to Kerans to have been resurrected from its own sewers. . . . The once translucent threshold of the womb had vanished, its place taken by the gateway to a sewer.
>
> (pp. 126 and 127)

Vagina and anus are here united in the cloaca of the city-mother (pp. 53 and 61). Drowned London had reverted from modern city to primeval swamp and now redrained London reverts from primeval swamp to post-modern slime and sewer covered in silt.

In accordance with the fundamental structuring principle of the novel whereby inner and outer are analogous, if not homologous, the drained lagoon affects Kerans profoundly and 'he began to sink rapidly into a state of dulled inertia, from which he tried helplessly to rouse himself. Dimly he realised that the lagoon had represented a complex of neuronic needs that were impossible to satisfy by any other means' (p. 129). Kerans experiences a sort of post-natal depression, not as mother, but as someone who has just been (re-)born. As a result, he becomes a kind of wetlands conservationist, though hardly an ecofeminist, who is sensitive to the homology between the inner and outer landscapes and aware of the connections between the depths and the surfaces, the repressed and the expressed, the past and the present. Although he is now fascinated with and not horrified by the wetland as womb, he is still not interested in his mother, and the Body of the Mother. The mother (in two senses) is absent from his story, though it/she is virtually present, or has left traces, in the references to the protozoan monster and black water of the lagoons. Despite his transformation from imperialist cartographer to wetlands conservationist, Kerans is still a hero from whose life, whose story, his mother is largely absent. Kerans extends his wetlands conservationism to becoming an eco-saboteur who blows up the dykes which hold the water back and heads south 'a second Adam searching for the forgotten paradises of the reborn sun' (p. 175).

Ballard's work is important as an attempt to tell the story of the imperial city as colonizing wetlands at 'home and abroad' and to think the unthinkable future albeit as repetition of the past in the return of those repressed wetlands. For Ballard in *The Drowned World* the future is thinkable, but only as a return to the past. This return is not nostalgic in the sense of homesickness, but *unheimlich* or unhomely which is both horrifying and fascinating. In *The Drowned World*, with its images of parodies of life, new life is being born from the wetland as womb.

These images include the destruction, and rebirth, of the world by water in *The Drowned World.* But not just any water, or even water per se, and certainly not the individualist death by water of T. S. Eliot's tired, modernist and mythopoeically nostalgic wasteland. It is the death of Western civilization by wetlands, by the return of its endo- and exo-colonial repressed, and a death which is a parodic repetition of its birth from the uncanny. The future is thought, not as difference, but as parody of sameness, the future as parodic repetition of the past, both of the species and the individual. The future is thinkable, but only as repetition of or return to the past. There are two kinds of countervailing trajectories at work and in tension here: one entailing a return to the repressed via the uncanny uterus and the other involving the future as a repetition of the uterine past. Perhaps the richness and suggestiveness of Ballard's novel can be attributed in part to the way in which the two lines are used to figure each other.

Part of its fascination can perhaps also be attributed to its quantum interplay with time and space, and with the way they are used to figure each other with colonized places figured as palaeozoic past; post-colonial here and now as pre-colonized and palaeozoic then and there. *The Drowned World* transforms our own ostensibly post-colonial and post-modern present into the determinate, and repressed, past of the pre-colonial and uterine yet to come.

The uncanny and the historiographic metafictional novel

The alligator as a vehicle and vector for the uncanny and the underside of the city as an uncanny place come together in Thomas Pynchon's *V*, first published in 1963, the year after J. G. Ballard's *The Drowned World*. These episodes of Pynchon's *V* are based on the urban myth or tall story of alligators in the sewers of New York City. Yet for Bruce Hallenbeck (2013, p. 61) 'the tales of alligators in the New York City sewers have much more than a grain of truth to them' and he goes on to present these loaves of truth. They date from newspaper accounts of the 1930s and culminate in the late 1950s and early 1960s with Robert Daley's *The World Beneath the City* with a chapter on 'Alligators in the Sewers' (Daley, 1959, pp. 187–189) and ultimately with Thomas Pynchon's *V*. These stories first emerged in the 1930s when Teddy May was the Superintendent of Sewers, or the 'King of the Sewers' as Daley (1959, p. 174) calls him in a chapter devoted to him, as 'he reigned below ground like a king' (Daley, 1959, p. 175). May deposed the alligators, the king of tropical swamps and marshes, as king of the sewers in an act of regicide. May was also 'Mayor under the streets of New York' and the Bruneseau of New York as 'he knew every turn and joint of New York's 560 miles of sewers' (Daley, 1959, pp. 174–175) just as Bruneseau did of the sewers of Paris (see Giblett, 2016, chapter 3). Both men mapped the sewers of their respective cities. Daley (1959, p. 175) goes on to relate how May 'also had the only topographical map of the original island, showing where the marsh ground had been.' This sounds suspiciously like the British Headquarters' Map of c.1782–1783. As Bruneseau was Napoleon's inspector of sewers in Paris, it is fittingly ironic that May had a 'Napoleonesque manner' (Daley, 1959, p. 175).

May was initially sceptical about alligators in the sewers until he saw them for himself when, like Bruneseau, he went on a voyage of discovery as another 'Columbus of the cloaca' of the new world beneath the city. A few alligators were hunted down and shot by May's inspectors armed with .22 rifles and pistols in what Daley (1959, p. 189) calls 'possibly the most unusual hunting on earth, a veritable sewer safari.' This urban myth or tall story is the basis for the episodes of Thomas Pynchon's *V* in which Profane is employed as an underground hunter on sewer safari armed with 'a 12-gauge repeating shotgun' (Pynchon, 1963, p. 113). Pynchon satirises May as Zeitsuss who 'was aware that most hunters regard use of this weapon like anglers feel about dynamiting fish; but he was not looking for write-ups in *Field and Stream*' (p. 113). May's and his men's happy hunting ground was the battlefield against the alligators and streams of sewerage underground that they made their home.

The story goes that alligators had got into the sewers when:

> Kids all over Nueva York bought these little alligators for pets. Macy's was selling them for fifty cents, every child, it seemed, had to have one. But

> soon the children got bored with them. Some set them loose in the streets, but most flushed them down the toilets. And these had grown and reproduced, had fed off rats and sewage, so that now they moved big, blind, albino, all over the sewer system. Down there, God knew how many there were. Some had turned cannibal because in their neighbourhood the rats had all been eaten, or had fled in terror.
>
> (Pynchon, 1963, p. 43; compare with Daley, 1959, p. 188)

In a post-modern parodic reprise of Melville's *Moby Dick* in which the obsessive Captain Ahab hunts the white whale across the greatest ocean of the world so Benny Profane hunts a pinto alligator ('pale white, seaweed black') beneath the greatest city in the world (Pynchon, 1963, p. 111) in an inverted, if not subverted, carnivalesque play on the great American novel. Just as Melville captured the maritime metropolis of New York and evokes its affects for New Yorkers of his day, so Pynchon captures the subterranean metropolis of New York and evokes its affects for New Yorkers of his day. Just as Melville's Ishmael made a myth out of whale hunting, so Pynchon's Profane and Angel 'added detail, color' to the story of the alligators in the sewers and together they 'hammered together a myth' (Pynchon, 1963, p. 142). Like the myth of the white whale, the myth of the alligators in the sewers taps into unconscious fears and desires. *Moby Dick* is the great novel of 'Man versus nature' across the oceans of the earth; *V* is the great novel of 'Man versus cultural nature' inside the sewers of the city.

Profane pursues the pinto alligator and confronts it in 'a wide space like the nave of a church' beneath the city that mirrors the churches above. 'A phosphorescent light coming off the walls whose exact arrangement was indistinct' illuminates this underground nave. Profane is at a loss to explain the light. He speculates that 'sea water shines in the dark sometimes; in the wake of ship you see the same uncomfortable radiance. But not here. The alligator had turned to him.' Profane eventually kills the pinto alligator in a nave-like space which is also 'a bonecellar, a sepulchre.' The killing of the alligator is a kind of secular mass or communion performed in a sepulchral nave in which 'blood began to seep out amoebalike to form shifting patterns with the weak glow of the water' (Pynchon, 1963, pp. 122–123) in a kind of milky, semenal discharge reminiscent of the sperm whales slaughtered for their blubber and oil in *Moby Dick*.

This episode haunts Profane who remembers it later and contemplates the inequitable exchange and the asymmetrical and non-reciprocal relationship between the hunter and the hunted (Pynchon, 1963, p. 146). The alligators give life because they give employment, but the hunting inspectors give death in exchange 'tit for tat' (Pynchon, 1963, p. 146). However, the hunted alligator can become the hunter and the hunting inspector can become the hunted as occurred 'when a gator turned and attacked' and Profane has to shoot it. Rather than the amoebalike-sacrament of the pinto alligator's demise that gives light and life in the underground nave lit by an uncomfortable radiance, the death of this alligator in the dark results in Profane 'standing by the headless corpse watching a steady stream of sewage wash its life blood out to one of the rivers'

(Pynchon, 1963, p. 147) in a secular and profane black mass in the sewers of the city. Profane is profane by name and nature.

The New York sewer in *V*, as Brian Jarvis (1998, p. 58) puts it, is 'portrayed allegorically as an underworld, vernacular geography of secret spaces and 'sewer stories' which go beyond truth, falsity and the rationalism of the alienating cityscape above.' Pynchon's counter geography of the world beneath the city in the sewers subverts the world of the city above and taps into the unconscious of the city. Unlike Bruneseau and May who were 'a Columbus of the cloaca' of Paris and New York respectively, Pynchon is the Columbus of the unconscious of the city who accepted Thoreau's invitation in *Walden* to 'be a Columbus to whole new continents and worlds within you, opening new channels, not of trade, but of thought' (Thoreau, 1982, p. 560). Pynchon is the Columbus of the world beneath the city, the grotesque lower urban stratum.

Pynchon taps into and operates in the register of the uncanny. For Freud and Pynchon alligators (and crocodiles for Freud as well) are vehicles and vectors for the uncanny. The uncanny for Freud was also applicable not only to the native quaking zone of the swamp where alligators and crocodiles live, but also to the artificial quaking zone of the dark underside of the city which for him was an object, or more precisely abject, of horror and fascination as we saw in both instances in previous chapters and as it is with Pynchon in *V* in a lighter, more playful vein. Pynchon's *V* combines both aspects of Freud's uncanny with the alligator in the sewer emerging as a figure for the return of and to the urban repressed.

References

Annear, R. 2014. *Bearbrass: Imagining Early Melbourne.* Melbourne: Black Inc.

Arnold, J. 1983. *The Imagined City: Melbourne in the Mind of Its Writers.* North Sydney: George Allen and Unwin.

Ballard, J. 1983. *The Drowned World.* London: Dent. First published 1962.

Beal, T. 2002. *Religion and Its Monsters.* New York: Routledge.

Benjamin, W. 1999. *The Arcades Project.* H. Eiland, trans. Cambridge, MA: Belknap Press of Harvard University Press.

Benjamin, W. 2014. *Radio Benjamin.* J. Lutes and others, trans. L. Rosenthal, ed. London: Verso

Bunyan, J. 2008. *The Pilgrim's Progress.* R. Pooley, ed. London: Penguin. First published 1678.

Cannon, M. 1976. *The Land Boomers.* New Illustrated Edition. Melbourne: Thomas Nelson.

Cunningham, S. 2011. *Melbourne.* Sydney: New South.

Daley, R. 1959. *The World Beneath the City*. Philadelphia: J. B Lippincott.

Freud, S. 2003. *The Uncanny*. D. Mclintock, trans. London: Penguin. First published 1919.

Geist, J. 1983. *Arcades: The History of a Building Type.* Cambridge, MA: The Massachusetts Institute of Technology.

Giblett, R. 1996. *Postmodern Wetlands: Culture History Ecology*. Edinburgh: Edinburgh University Press.

Giblett, R. 2008a. *The Body of Nature and Culture.* Basingstoke: Palgrave Macmillan.

Giblett, R. 2008b. *Sublime Communication Technologies.* Basingstoke: Palgrave Macmillan.

Giblett, R. 2011. *People and Places of Nature and Culture.* Bristol: Intellect Books.
Giblett, R. 2016. *Cities and Wetlands: The Return of the Repressed in Nature and Culture.* London: Bloomsbury.
Giblett, R. 2018. *Environmental Humanities and Theologies: Ecoculture, Literature and the Bible.* London: Routledge.
Grann, D. 2017. *Killers of the Flower Moon: Oil, Money, Murder and the Birth of the FBI.* London: Simon and Schuster.
Grimwade, S., ed. 2009. *Literary Melbourne: A Celebration of Writing and Ideas.* Prahran: Hardie Grant.
Hallenbeck, B. G. 2013. *Monsters of New York: Mysterious Creatures in the Empire State.* Mechanicsburg: Stackpole Books.
Hoffmann, E. 1991. The Mines of Falun. In: *Spells of Enchantment: The Wondrous Fairy Tales of Western Culture.* J. Zipes, ed. L. Kent and E. Knight, trans. New York: Viking Penguin, pp. 304–324. First published 1819.
Hume, F. 1999. *The Mystery of a Hansom Cab.* Melbourne: Text Publishing. First published 1886.
Jarvis, B. 1998. *Postmodern Cartographies: The Geographical Imagination in Contemporary American Culture.* New York: St. Martin's Press.
Lord Dunsany. 1908. The Kith of the Elf-Folk. In: *The Sword of Welleran and Other Stories.* London: George Allen & Sons, pp. 31–45.
Lovecraft, H. 2001. The Music of Erich Zann. In: *The Thing on the Doorstep and Other Weird Stories.* S. T. Joshi, ed. New York: Penguin, pp. 45–52. First published in 1922.
Lovecraft, H. 2004. The Moon-Bog. In: *The Dreams in the Witch House and Other Weird Stories.* S. T. Joshi, ed. New York: Penguin, pp. 42–49. First published in 1926.
Otto, K. 2005. *Yarra: A Diverting History.* Melbourne: Text Publishing.
Pynchon, T. 1963. *V.* New York: Random House.
Stoker, B. 2011. *Dracula.* Oxford: Oxford University Press. First published 1897.
Sussex, L. 2015. *Blockbuster! Fergus Hume and the Mystery of a Hansom Cab.* Melbourne: Text Publishing.
Thoreau, H. 1982. *The Portable Thoreau.* C. Bode, ed. New York: Viking Penguin.
Vidler, A. 1992. *The Architectural Uncanny: Essays in the Modern Unhomely.* Cambridge, MA: MIT Press.

8 The uncanny Commonwealth of Christianity

The legend of Saint George killing the evil dragon is well known. It's in books and statues and on coins. Throughout the Commonwealth, institutions and streets are named 'St Georges.' St George is the patron saint of England, the Church of England and the Scouts. He is a monumental figure of monotheistic agricultural and pastoral Christianity while the dragon is a monstrous creature of polytheism who inhabits wetlands, both of which need to be destroyed to establish the Christian religion in the Christian nation, the pastoral dryland in the pagan wetland and the Christian colony in the Christian empire. The legend is a foundation myth of English nation-building and British colonization and empire-building, including the colonization of nature in general and of what I call 'placial discrimination' and 'placism' against such places as wetlands (see Giblett, 2016, chapter 3, 2018b, chapter 2).

This chapter calls for the decolonization of dragons as evil monsters and their remythification as divine monsters. It begins with a brief recount of the legend and goes on to consider the dragon in British culture and history. Recently I have undertaken the decolonization of dragons by retelling legends and stories about them from an environmentally and animal friendly point of view (Giblett, 2018a, 2018c). The present chapter then draws on the theory of the history, politics and role of the Commonwealth as outlined by Cultural Studies pioneer Stuart Hall and his co-author Noelle Goldman (Goldman and Hall, 1987), by the now largely forgotten Patrick Gordon Walker (1962) in his tome *The Commonwealth* and recently by Philip Murphy (2018) in *The Empire's New Clothes: The Myth of the Commonwealth*, all of whom consider religion and race, but none of whom consider Christianity as central to the projects of colonization of nature and places such as wetlands and decolonization of nature and wetlands.

Although Murphy sees the Commonwealth as a 'pale Protestant replica' of the Roman Catholics, as 'a sort of global secular religion' and acknowledges 'a new politics of intersectionality' around 'class, race, religion and gender,' in which there is 'a growing intolerance of 'racial discrimination' and concerns about immigration, none of these aspects are discussed in any detail as Murphy is an empiricist Commonwealth historian and policy wonk (Murphy, 2018, pp. 15, 58, 107, 140, 161, 195, 202 and 229). His book is based on oral history interviews with key informants who set the agenda for the discussion and who tend to give the answers to questions that are sought. The limitations of this insider ethnographic style of research are not acknowledged.

The Commonwealth of Christianity with a shared language and a common tradition and institution in the Christian churches is the handmaid to the British political and economic empire with their corporations currently colonizing the world, the planet and beyond in the Commonwealth with the other three evil empires of China, Russia and the United States colonizing nature. The Commonwealth is also a Commonwealth of counter-culture of writers and other cultural workers who work in solidarity with each other against the four evil empires and for the decolonization of people, places and the planet.

This counter-cultural Commonwealth is completely ignored or overlooked by Murphy (2018) who does not define the Commonwealth and refers to it repeatedly as an organization, not as embracing and facilitating a network of interest groups, such as the various Associations for Commonwealth Literature and Language based in many Commonwealth and non-Commonwealth countries. The various ACLALS are not mentioned or discussed at all. The 'myth of the Commonwealth' that Murphy (2018, p. 3) demythologizes is largely a British myth in which the 'Empire continues to live in the [British] public imagination.' It also lives in different ways in the non-British public imagination of many Commonwealth *not* as 'Empire' as the ACLALS show and live. Murphy (2018, p. 6) does indicate as much when he suggests that 'today's Commonwealth [is] an organization which tried to define itself in terms of virtually everything the British Empire was not,' but that is not the main thrust of his book. It is primarily about the *British* myth of the Commonwealth as Murphy (2018, p. 226) acknowledges in his conclusion.

In conclusion, this chapter turns to the Queen's Commonwealth Canopy in which Commonwealth countries commit to conserving forests and woodlands and/or planting trees (Murphy, 2018, p. 102). Begun in 2015, nearly half the countries of the Commonwealth have signed up to it and instituted conservation or rehabilitation projects. A greater challenge for the Commonwealth would be to conserve or rehabilitate wetlands by planting native vegetation in what could be called the Queen's Commonwealth Wetlands. The Commonwealth is well-positioned to initiate such a programme as 25% of the world's wetlands are in Canada (Giblett, 2014) and Australia has lost 75% of its pre-contact wetlands (Giblett and Webb, 1996).

Colonization is just as much about the colonization of nature and places as it is about the colonization of people. It is just as much about draining of swamps and building railways through the bush as it is about a native population that it is politically and economically oppressed and exploited. Indeed, for Franz Fanon, the pioneer theorist of decolonization, they are one and the same thing:

> Hostile nature, obstinate and fundamentally rebellious, is in fact represented in the colonies by the bush, by mosquitoes [from swamps], natives and fever [from mosquito bites], and colonization is a success when all this indocile

> nature has finally been tamed. Railways across the bush, the draining of swamps and a native population which is non-existent politically and economically are in fact one and the same thing.
>
> (Fanon, 1967, p. 201)

Colonization is also just as much about religion and the triumph of monotheism over polytheism, of colonizing monotheistic Christianity over polytheistic local religions of the colonized. Decolonization is not just about the departure of the colonizing power and the instituting of 'native' self-rule, but also about the decolonization of 'nature.' Far from living in the era of 'post-colonialism' trumpeted by many writers, we are living in the era of neo-colonialism as decolonization of nature has not occurred. The colonization of nature lives on.

Nor has decolonization of place occurred in the spatial hierarchy of the colonial settlement. The archetypal colonial city for Fanon (1967, p. 30) was divided between the Settler's Town basking on the hillsides and the Native Town 'wallowing in the mire' as it does on the mud flats of the Ganges in Chandrapore and as depicted in E. M. Forster's *A Passage to India* (Forster, 1936, p. 9). Melbourne is a classic case in point as it had a similar spatial hierarchy with the upper crust basking on the salubrious hillside of Toorak while the working classes wallowed in the miry slums of Collingwood on the other, lower side of the Yarra River and the 'Natives' wallowed in the mire of Fitzroy and Treasury Gardens and in the mud flats of the Yarra River (see Giblett, forthcoming, chapter 8). Placial discrimination is exercised by the people of the upper place over the lower place and its people.

Nor has decolonization of the mind occurred either if the prevalence of Christianity and its institutions and legendary figures such as St George in many former colonies is anything to go by. Decolonization of the mind is also not just about decolonizing language and literature, but also about decolonizing religion. In *Decolonising the Mind* Kenyan writer Ngugi wa Thongi bemoans the fact that 'the Christian bible is available in unlimited quantities in even the tiniest African language.' Even worse for him is the fact that:

> the most zealous of European missionaries who believed in rescuing Africa from itself, even from the paganism of its languages, were nevertheless masters of African languages, which they often reduced to writing. The European missionary believed too much in his mission of conquest not to communicate it in the languages most readily available to the people.
>
> (Ngugi, 1986, p. 26)

Ngugi goes on to bemoan the fact that the African writer at that time did not communicate in the languages most readily available to the people, but in European languages, and did not believe in or practise decolonization sufficiently to communicate in his or her 'mother tongue.'

The colonial mission of conquest led to the triumph not only of one power and language over another, but also of monotheistic Christianity over polytheistic paganism. None of these missions or triumphs could have taken place without the other, nor without the translation of the Bible and the communication of its message in 'native' languages, nor without the triumph of Christian legends, such as St George and the Dragon, over pagan fables and tales.

St George and the Dragon

The dragon is a legendary figure in many cultures. In English culture and literatures in English the dragon is traditionally a flying, evil, masculine monster who wreaks death and destruction with his fiery breath. The dragon in *Beowulf* and Smaug in Tolkien's *The Hobbit* is a monstrous serpent who lives underground, flies at night or twilight, and embodies and exhales fire and/or venom (see Giblett, 2018b). The dragon is portrayed in both stories as a creature of primeval fiery chaos and as an earth monster who is a terrible hoarder of precious minerals and metals. The dragon in the legend of St George and Satan in Milton's *Paradise Lost* is also a monstrous serpent. The dragon is portrayed as a creature of primeval watery chaos and as a marsh monster who is a horrible creature of slimy swamps (see Giblett, 2018b).

I read and critique the portrayal of dragons as evil monsters. I argue that earth dragons are keepers and guardians of minerals and metals in the earth. Humans steal minerals and metals from the earth/dragon, portray earth dragons as robbers and evil monsters, heap their guilt for doing both on to them as scapegoats and then kill them and waste the earth to expiate their guilt. Earth dragons on this view are projections of human guilt for robbing the riches of the earth. A more environmentally and animal friendly view of earth dragons would see them as manifestations of the precious metals and minerals of the earth to live with and use sacrally and symbiotically, rather than to kill (Giblett, 2018c).

Earth dragons wreaking fiery vengeance on humans for stealing from the earth can also be read as a climate change allegory. More intense bush and forest fires in the age of global warming are the earth's vengeance for greedy mining and carboniferous capitalism. Living more sacrally and symbiotically with dragons as creatures of fire and the earth as combustible means drastically reducing greenhouse gas emissions.

Swamp dragons function culturally and environmentally in a similar way to the earth dragons. Humans steal new life from the fertile swamps/dragon, portray swamp dragons as evil monsters, heap their guilt for doing both on to them as scapegoats and then kill them and drain swamps to expiate their guilt. Swamp dragons on this view are projections of human guilt for living parasitically on the good things of the fertile swamps. A more environmentally and animal friendly view of swamp dragons would see them as manifestations of the fertility of the earth to live with and use sacrally and symbiotically, rather than to kill (see Giblett, 2018a).

Perhaps the most famous dragon of all in English culture and its settler diasporas is the swamp dragon of the legend of St George and the Dragon. In the eleventh and twelfth centuries of the Common Era the legend of St George and the Dragon started to circulate widely in Europe. Its success in England was partly attributable to *Beowulf* and its dragon. In one version of the St George legend 'a terrifying Dragon' is an inhabitant of a lake and an instrument of God's wrath. God sent the dragon 'which devoured the inhabitants' of the nearby city ruled by 'a godless emperor' (Collins, 2012, p. 90). St George kills the 'huge and awful' dragon with his sword.

St George is depicted killing the dragon with his sword in probably the most famous depiction of them all, Benedetto Pistrucci's design of St George and the Dragon on many mintings of British gold sovereigns and half-sovereigns dating from 1817. Collins (2012, p. 203) describes this design in his exhaustive catalogue of all things related to St George and the Dragon as 'one of the most beautiful designs ever to appear on a coin.' It may be a beautiful design, but it depicts the gruesome and horrible slaying of the dragon in an act of gross animal cruelty. The beauty of the design hardly mitigates or overcomes the ugliness of the dragon-slaying.

For Ingersoll (2014, p. 194) writing in the 1920s the two main protagonists of the legend are 'the proper, most eminent Saint George, and his most celebrated and distinguished of all Dragons.' Unlike the dragon of *Beowulf* who is an earth monster, the dragon of the legend of St George is a water monster, or more precisely a marsh monster, like Grendel and his mother in *Beowulf*, and/or a swamp serpent. The dragon in general for Ingersoll (2014, p. 15) has 'only one definable characteristic – association with and control of water.' This is certainly not the case with the dragon of *Beowulf* who has a number of defining characteristics, such as living in the earth and association with, and control of, fire. In Ingersoll's summary of the St George legend 'a loathsome monster' inhabits 'a pond' of 'dark water.' Like the dragon of Beowulf, 'the exhalations of its breath poisoned all who came near.' St George meets 'the horrible fiend' and 'transfixed the beast' with his lance (Ingersoll, 2014, pp. 194–195). Most depictions of St George and the Dragon show him killing the dragon with his lance.

In the thirteenth century CE *The Golden Legend* for Collins (2012, p. 91) 'popularized the legend of George and the Dragon in the West. [. . .] The story may have been particularly well-received in England because of the legend of Beowulf in Anglo-Saxon literature' dating from the eighth century containing the English Ur-story of a dragon. Caxton published an English translation of *The Golden Legend* in 1483. It describes 'a stagne or a ponde lyke a see wherein was a Dragon whyche envenymed all the contre' (cited by Collins, 2012, p. 93). 'Envenomed' is largely lost from English today (Robert Coover uses it in 'The Dead Queen'), but is certainly precise when considering the activities of the dragon of *Beowulf*. The dragon of St George follows suit. The dragon in *The Golden Legend* is not merely a water monster, but specifically a swamp serpent or marsh monster. This is certainly the case for Baden-Powell, the founder of the Boy Scouts, for whom the dragon 'lived in a swamp' (cited by Collins, 2012, p. 128). Similarly Sabine Baring-Gould's *Lives of the Saints* of 1872 recounts the legend in which 'a pond [is] infested by a monster' and how St George sees 'the monster rising from the marsh' (cited by Collins, 2012, p. 159).

The dragon in Christianity, Collins (2012, p. 92) concludes, 'symbolized the Devil, heresy, paganism and ultimately anything destructive of [Christian?] society. [. . .] Many a Dragon has the curled and twisted tail of a serpent, suggestive of evil as old as Eden.' The dragon in Boehm's statue of St George and the Dragon outside the State Library of Victoria in Melbourne has a forked tongue, which is certainly suggestive of a serpent and of deception (see cover photo of Giblett, 2018a). Whether this is also suggestive of evil as old as Eden or not is another question. Geoffrey Serle (1971, p. 287) might concur with Collins when he argued that this statue signifies 'the triumph of good over evil.' Yet the dragon as a creature of primeval chaos as depicted in *Genesis* 1:2 may be older than Eden and not suggestive of evil at all. A twisted tail and forked tongue are suggestive of a serpent, but are not necessarily suggestive of evil, as old as Eden or not, and only with the hindsight of the biblical story of the Fall.

Contrary to Serle, I would argue that the statues and coins of St George killing the dragon signify the triumph of monotheism over polytheism. They are the first of the Ten Commandments in a graven image: 'Thou shalt have no other [polytheistic] gods before me,' the singular, monotheistic, capital 'G' God. The second commandment interdicts the worshipping of graven images. It does not interdict the forming, non-worshipping and using for instructive purposes of graven images in statues and coins of God's opponents, such as dragons, and saints, such as St George.

Dragons are relatedly figures of endo- and exo-colonialism. They are mobilized and deployed for internal colonialization within the home country of the colonizing power and for external colonization of the conquered lands in the colonies. Swamp dragons are used to justify the enclosure of the commons and the drainage of wetlands. Earth dragons are used to justify the robbery of the mineral wealth of the earth. In English literature the dragon is portrayed as a creature either of primeval fiery chaos and as an earth monster or of primeval watery chaos and as a marsh monster. Recent retellings of the legend of St George and the Dragon and of the story of Beowulf and the dragon from an environmentally and animal friendly point of view decolonize dragons (see Giblett, 2018a, 2018c).

Saint Margaret and the Dragon

Another dragon and another Saint concerns the legend of St Margaret and the Dragon. St Margaret was canonized because she refused to marry a pagan priest, suffered various ordeals with a dragon and was eventually martyred for her trouble. She was imprisoned for her crime and consumed in her prison cell by a dragon who 'burst apart' when she made the sign of the cross and she escape unharmed (Clayton and Magennis, 1994, pp. 5 and 123), or she 'managed to escape by using a cross to irritate the monster's insides' (Brown, 2018). St Margaret became a cult figure which the Vatican banned in 1969 (Clayton and Magennis, 1994, p. 3). In the twelfth century CE she became the patron saint of childbirth (Clayton and Magennis, 1994, p. 4), perhaps because she was reborn as a saint from the dragon. As Clayton and Magennis (1994, pp. 4 and 36) put it in their scholarly monograph on the Anglo-Saxon lives of St Margaret, 'the grotesque and appalling image of Margaret being swallowed by the dragon [. . .] with graphic attention to detail is a definitive feature of iconographic representations of St Margaret.' The most famous of these is Titian's painting of circa 1559 CE which shows the dragon with St Margaret carrying a small crucifix (Brown, 2018).

In one version of the life of St Margaret the dragon is described as:

> a most terrifying dragon of many different colours. His hair and beard appeared golden, and his teeth were just like cut iron, and his eyes shone like strange gems, and from his nose came a great deal of smoke, and his tongue breathe out and he caused a tremendous stench in the prison. Then he raised up and hissed with a loud sound. Then there was a great light in the dark prison from the fire which came out from the dragon's mouth.
>
> (Clayton and Magennis, 1994, p. 123)

The terrifying dragon, or dreadful and fearsome in another version of the life of St Margaret, is an orally sadistic monster with sharp teeth whose mouth also projects consuming fire (Clayton and Magennis, 1994, p. 205). He is also a monstrous serpent who hisses loudly and emits loathsome smells.

In another version of the life of St Margaret:

> A horrible devil entered [Margaret's prison cell]. His name was Rufus and he was very large, with the appearance of a dragon, and he was all spotted like a snake. And a glare emanated from his teeth, as if from a bright sword, and it was as if a flame of fire came from his eyes and smoke and an immense great fire from his nostrils, and his three tongues lay around his neck.
>
> (Clayton and Magennis, 1994, p. 163)

Rufus the dragon is a red-haired devil. He is such an orally sadistic monster that he has three tongues all the better to eat you with. He is also a creature of fire who consumes with fire as in the previous version. The devil-dragon 'wanted to devour the girl,' but she 'made the sign of the cross on her forehead and so protected herself thoroughly against the dragon' (Clayton and Magennis, 1994, p. 163). He is a vehicle and vector for the uncanny as a return to and of the pre-Christian pagan and polytheistic past.

The Commonwealth according to recent writers

The colonization of nature and the conquest of monotheism over polytheism are aspects of the Commonwealth, colonization and decolonization not on the radar of Patrick Walker who was the Secretary of State for the Commonwealth from 1950 to 1951 and Under-Secretary of State for the Commonwealth from

1948 to 1950. The promulgating of Christianity in Africa was certainly tied up with colonization. For Walker (1962, p. 62) converting the people of West Africa to Christianity was instituted by leaders of the British anti-slavery movement to stop the slave trade by turning to 'more radical cures' than interfering with the trade. In other words, they wanted to treat the cause of the disease, rather than the symptoms. Walker goes on to point out that 'missionary activity began in the first half of the nineteenth century; but it made little headway until the European power established colonies in West Africa.' Colonization, in other words, road on the coat tails of Christianity, but Christianity made few inroads until colonies were established to provide the context and support in which Christianity could take root. Christianity and colonization eventually worked together. Even then and there it was far from plain sailing as Murphy (2018, p. 139) points out 'missionaries were viewed with some suspicion [. . .] in Nigeria [where] British control was firmly based on the collaboration of a series of Muslim emirates.' Christianity and colonization were mutually reinforcing, one providing the moral (and monotheistic) message, the other the institutions of governance and education in English that made the promulgation of Christianity possible which, in turn, reinforced colonization.

The West Indies is a case in point too, as Walker (1962, p. 75) simplistically puts it, as 'its people were overwhelmingly of West African origin: but, through the conditions of slavery, they had lost their tribal unities of language and culture. They became English-speaking and Christian.' Yet not all West Indians became Christians, as Walker goes on to point out in the case of West Indians of 'East Indian' origin who retained their polytheistic Hindu religion. Many West Indians of West African origin retained their West African polytheistic cultural beliefs and practices in their diaspora.

More recent writers on the Commonwealth are more sensitive than Walker to its cultural politics around language, religion and ritual, though without addressing the triumph of monotheism over polytheism. Noelle Goldman and Stuart Hall, the Cultural Studies pioneer and West Indian immigrant to the UK, collaborated on the book of photographs for the Commonwealth Photography Award in 1987 by contributing their own essays to accompany the long-listed entries. Hall (Goldman and Hall, 1987, p. 12) comments that 'the Award aimed to use the Commonwealth positively, as a sort of prism for contemporary relationships between metropolis and periphery, between the developed and developing worlds, across the divide between "North" and "South."' The Commonwealth in the north and south continues to be a refractory device for producing solidarity among counter-cultures against the four evil empires of the north.

These contemporary relationships also include relationships with and between religions and with what Goldman later calls 'the ecology of the fragile planet that is Earth.' She is much more attuned to the politics of religion and ecology than Walker and Murphy. She acknowledges that 'religion has its structural relations to the state. Most religions are major political and social institutions, which, through ritual, control our lives' (Goldman and Hall, 1987, p. 91). Christianity is a case in point of a major political and social institution which directs lives, not just human ones but also non-human ones such as plants and animals.

Ritual for Goldman is the overarching cultural category of which religion is a subset. Ritual for Goldman punctuates the flow of time in everyday life and:

> lends a special meaning and significance to gatherings, celebrations, carnival, and remembrance. [. . .] On a more sinister level ritual activity can also be used to try to exert control over what is threatening and unknown or to control us through particular social relations.
>
> (Goldman and Hall, 1987, p. 91)

Religion plays a role in both aspects of ritual, yet for Goldman 'religious and spiritual experience has a more universal level of existence which is outside time and space.' For Goldman and Hall (1987, p. 91) through this 'more universal level of awareness [. . .] we are connected to each other and the ecology of the fragile planet that is Earth.' Religious and spiritual experience seems to be implicitly for Goldman the way of making this connection and raising this awareness. Yet she does not specify which religions (whether monotheistic or polytheistic) and what sort of spiritual experience might enable and nurture this connection and awareness. Some monotheistic religions, such as Christianity, have a problematic and fraught relationship with 'the fragile planet that is Earth' (see Giblett, 2018b), while other polytheistic religions and spiritualities, such as Taoism, make the connection to, and raise awareness about, 'the fragile planet that is Earth' (see Giblett, 2008, chapters 10 and 11).

The Commonwealth as a political and cultural institution also has its own rituals. Rituals are religious or secular, or a mix of both. Patrick Murphy's *The Empire's New Clothes: The Myth of the Commonwealth* was published to coincide with the Commonwealth Heads of Government Meeting (CHOGM) in London in April 2018. Coincidentally, publication of this book also occurred at the same time as the Commonwealth Games were being held on the Gold Coast in Australia. Walker (1962) does not mention (Murphy (2018, pp. 44 and 228) does) this abiding ritual and institution of the Commonwealth which began in 1930 and which occurs every four years, two years apart from the Olympic Games. Both events are rituals with religious and secular overtones, especially with the CHOGM meeting opening in Westminster Abbey, the seat of the Church of England, with monotheistic overtones that asserted the alignment of the Commonwealth states with Christianity and the dominance of Christianity officially in the Commonwealth as one of its unifying factors. Christianity is the most widely distributed religion among the countries of the Commonwealth.

The CHOGM of April 2018 and the unsurprising appointment of Prince Charles during it as the incoming Head of the Commonwealth by the current Head and the Head of the Church of England, his mother the Queen, caused pause for reflection about the nature and future of the Commonwealth in the world which, by this very act, smacks of a hereditary royalist, monotheistic religious and imperialist institution and traditions. It also prompted Afua Hirsch

(2018, p. 48) writing in *The Guardian Weekly* to ask, 'What is the Commonwealth if not the British Empire 2.0?' Of Prince Charles she writes:

> Who else other than the heir to the British throne, after all, could be better qualified to lead the contemporary manifestation of the British empire? It would just be so much easier if all concerned simply admitted this reality: the Commonwealth is a vessel of former colonies with the former imperial master at its helm. Or, as I like to call it, Empire 2.0.

The Commonwealth is a ship of states with the helm manned by an adherent of one monotheistic religion, or at least his mother is according to Murphy (2018, p. 99). Hirsch goes on to enumerate the economic imbalances of the flows of wealth in the so-called 'Commonwealth' with British companies controlling $1trillion of Africa's key resources while 'Africa loses around $40billion more each year than it receives in aid, loans and remittances.' The wealth of the Commonwealth of nations is not shared in common among its member nation-states. The Commonwealth is not common-wealth. Without giving specific figures, Murphy (2018, p. 129) acknowledges the 'unequal trading relations and extraction of mineral resources designed to benefit Europe rather than the developing countries concerned.' The Commonwealth is not just an organization of states, but a network of countries through which corporations channel resources and capital to their benefit and the benefit of their shareholders along the old routes of Empire back to the centre and seat of power in the UK.

Hirsch also highlights the immigration injustices in the flows of people of colour in the so-called 'Commonwealth' with:

> The Caribbean heads of state wanting to know why Britain, having invited their strong, talented and motivated citizens to this country in the 1960s [such as Stuart Hall], has been seeking to deport their children in the 2010s.

Race for Hirsch:

> has been there from the very beginning [in] the original laws restricting immigration from the Commonwealth [. . .] introduced in 1961 which were 'intended to operate on coloured people almost exclusively.' In 1973, the [Conservative] government [of Edward Heath] introduced measures to mitigate the impact [of this legislation] on white migration from the Commonwealth, [principally] from Australia, Canada and New Zealand. This led to their description by the Labour party in opposition at the time as 'cruel and brutal anti-colour legislation.' The treatment of applicants from the non-white Commonwealth has, immigration law experts agree, 'been markedly different.'

The people of the Commonwealth are not treated in common with each other by the former imperial master, nor are places, such as wetlands with many being drained or filled as part of the project colonization, so as to create spaces for colonial settlements and cities, in the process destroying the places that were there at contact.

The Commonwealth photography award of 1987 acknowledged that 'the world is made up of millions of people of different races, colours and creeds' (Goldman and Hall, 1987, p. 33). Creeds here stands in for religion. Like rituals, creeds are an overarching category. Whereas rituals are practices, creeds are beliefs, either explicit in written form or implicit in attitudes and values held by people. The Commonwealth even has a creed that Murphy (2018, p. 16) calls 'a sort of profane version of the Nicene Creed' that includes affirmation of the worth and dignity of every person and of 'our respect for nature, and that we will be stewards of the earth by caring for every part of it, and for it as a whole.' This creed, according to Murphy (2018, p. 16), is 'rather awkwardly recited by the congregation at the annual Commonwealth Day Observance service in Westminster Abbey.' The awkwardness might be a result of the creed's litany of motherhood statements and its outmoded biblical language, especially its allusion to the King James Versions of *Genesis* 1: 28 and *Psalm* 24: 1 around 'stewardship,' which is now seen as legitimating mastery over the earth, rather than nurturing mutuality with it (see Giblett, 2018b, chapter 5).

An entire section of the Commonwealth photography award of 1987 and the book is devoted to the theme of the people of the Commonwealth of these races, colours and creeds (Goldman and Hall, 1987, pp. 34–61). People also appear in all the other photographs. There are no wilderness or landscape photographs of places without people. The award is devoted to people of the Commonwealth in their places, and not to the places themselves without people. The photographs do not display or promote people engaged in activities that show respect for 'the fragile planet that is Earth' (see Giblett and Tolonen, 2012, especially chapter 13, 'Photography for Environmental Sustainability').

In his 1962 book on the Commonwealth Walker (1962, pp. 344–361) devotes a chapter to 'Race and Colour' in which he concentrates on 'racial problems' in the Commonwealth countries but largely excludes the United Kingdom (is it not a Commonwealth country?), except for this brief dismissal (the UK does not rate an index entry under the category of 'Racial problems' (p. 403) while the other major Commonwealth countries do):

> In the United Kingdom a considerable inflow in the 1950s from India, Pakistan, West Africa and especially the West Indies led to active discussion about a change in Britain traditional policy in order to control the immigration of Commonwealth citizens.
>
> (Walker, 1962, p. 345)

Of course, the Commonwealth citizens of these two nations and regions are principally coloured. He has nothing to say about the immigration (permanent or temporary) of Commonwealth citizens from Australia, Canada and New

Zealand. Walker (1962, p. 381) concludes his book by calling for 'the ending of race-discrimination' and enunciating other noble aims for and within the Commonwealth, but that day has far from arrived. The racial repressed of the Commonwealth returns in the writing about the Commonwealth while the religious repressed remains a more shadowy and vague presence/absence.

Similarly, the ending of place discrimination against wetlands has far from arrived with their continued destruction and loss. The natural repressed of the Commonwealth remains a shadowy and vague presence/absence in placial discrimination against denigrated, literally blackened, places such as wetlands. The Queen's Commonwealth Canopy begun in 2015 is a program to which about half the Commonwealth countries have signed up to and committed to conserving and/or rehabilitating forests and woodlands, often by planting trees (Murphy, 2018, p. 102). A greater challenge for the Commonwealth would be to institute the Queen's Commonwealth Wetlands in which Commonwealth countries would commit to conserving and/or rehabilitating wetlands, often by planting native vegetation. Doing so would not only end placial discrimination against wetlands, but also bring nature, religion and the repressed back together in decolonization and resacralization of the earth and its creatures.

References

Brown, M. 2018. Titian Painting Given to Charles I's Plumber Goes Up for Sale. *The Guardian*, 22 January. Available online at: www.theguardian.com/artanddesign/2018/jan/22/titian-painting-king-charles-i-plumber-sale-auction

Clayton, M. and H. Magennis. 1994. *The Old English Lives of St Margaret.* Cambridge: Cambridge University Press.

Collins, M. 2012. *St George and the Dragons: The Making of English Identity.* London: CreateSpace.

Fanon, F. 1967. *The Wretched of the Earth.* C. Farrington, trans. Harmondsworth: Penguin.

Forster, E. 1936. *A Passage to India.* Harmondsworth: Penguin.

Giblett, R. 2008. *The Body of Nature and Culture.* Basingstoke: Palgrave Macmillan.

Giblett, R. 2014. *Canadian Wetlands: Places and People.* Bristol: Intellect Books.

Giblett, R. 2016. *Cities and Wetlands: The Return of the Repressed in Nature and Culture.* London: Bloomsbury.

Giblett, R. 2018a. *The Dragon and Saint George: A Fairy Tale Novella.* Cambridge: Vanguard Press.

Giblett, R. 2018b. *Environmental Humanities and Theologies: Ecoculture, Literature and the Bible.* London: Routledge.

Giblett, R. 2018c. *Tales of Two Dragons.* London: Austin Macauley.

Giblett, R. forthcoming. *Modern Melbourne: City and Site of Nature and Culture.* Bristol: Intellect Books.

Giblett, R. and J. Tolonen. 2012. *Photography and Landscape.* Bristol: Intellect Books.

Giblett, R. and H. Webb, eds. 1996. *Western Australian Wetlands: The Kimberley and South-West.* Perth: Black Swan Press/Wetlands Conservation Society.

Goldman, N. and S. Hall. 1987. *Pictures of Everyday Life: People, Places and Cultures of the Commonwealth.* London: Comedia.

Hirsch, A. 2018. What Is the Commonwealth If Not the British Empire 2.0? *The Guardian Weekly*, 27 April, p. 48. Also available online at: www.theguardian.com/commentisfree/2018/apr/17/commonwealth-british-empire-britain-black-brown-people

Ingersoll, E. 2014. *Dragons and Dragon Lore.* New York: Cosimo Classics. First published 1928.

Murphy, P. 2018. *The Empire's New Clothes: The Myth of the Commonwealth.* London: C Hurst.

Ngugi wa Thongi. 1986. *Decolonising the Mind: The Politics of Language in African Literature.* Oxford: James Currey.

Serle, G. 1971. *The Rush to Be Rich: A History of the Colony of Victoria, 1883–1889.* Carlton: Melbourne University Press.

Walker, P. 1962. *The Commonwealth.* London: Secker & Warburg.

9 The living polytheism of the Fung Loy Kok Institute of Taoism/Taoist Tai Chi Society

Polytheism is alive and well in the Fung Loy Kok Institute of Taoism/Taoist Tai Chi Society (hereafter 'FLK' or 'the Society'). This not-for-profit organization is dedicated to bringing the beliefs, practices and benefits of the 'three religions of China' (Buddhism, Confucianism and Taoism) to all communities. One of the aims and objectives of the Society is to bring the riches of Chinese culture to others in cultural exchange. Founded in Canada in 1970 by Master Moy Lin-shin, the Society is active in over 20 countries and is the largest volunteer tai chi organization in the world. This chapter draws on the publications of the Society, including transcripts of talks by Master Moy and translations of Chinese texts, to present these beliefs and the benefits of practising the Taoist arts of health, including taijiquan and meditation.

The first major book publication of FLK is entitled *A Path of Dual Cultivation: Teachings of the Fung Loy Kok Institute of Taoism* (FLK, 2008). The first sentence of the chapter devoted to an overview of the 'Taoist Deities' introducing the third and final part of the book begins by stating that 'Taoism is polytheistic, which means that there are myriad deities, immortals, lords, spirits and so on in the Taoist pantheon' (FLK, 2008, p. 113). No mention is made here of 'religion.' Is Taoism a religion? Of course, it all depends on how you define 'religion.' The 'Preface' to this book addresses this question and gives a direct answer to it by stating that 'in many Eastern traditions, religion is seen more as a school of teaching than as a devotional faith with strict doctrines as is often seen in the West' (FLK, 2008, p. vi). Taoism is a 'school of teaching' that does not exclude other schools of teaching (or 'religions'). Taoism is the way of the way (or Tao). Taoism follows 'the Great Way (the Tao)' (FLK, 2008, p. 19). Similarly, Taoist deities are not gods in the sense of being separate beings but are embodiments of virtues or expressions of the body and mind. They are not without but within.

Taoism has 'integrated' Buddhism and Confucianism (pp. vi–vii). It has a pantheon (literally 'all the gods together'), like many other peoples who have a collection of gods, such as ancient Greece and Rome, and other religions, such as Hinduism. The Taoist pantheon is polytheistic. Taoism does not have 'strict doctrines,' but it does have virtues that enshrine values and guide behaviour derived mainly from Buddhism and Confucianism, such as 'the Eight Virtues' of filial piety, sibling harmony, dedication, trustworthiness, propriety, sacrifice, honour and sense of shame (the virtue you should have when you don't have the others) (FLK, 2008, pp. 19–29). Taoism is devotional with the study of scriptures and the practice of meditation, and it has faith, and hope and love. Compassion is one of its overriding virtues.

Compassion is embodied in the Buddhist Bodhisattva of compassion, Guan Yin, who was Master Moy's personally favourite deity. The story goes that he was a sickly child rejected by his mother at birth and brought up by his older sister who later took him to Taoist monks to try to get them to improve or regain his health. He promised Guan Yin that if he regained his health from practising the Taoist arts of health, he would dedicate the rest of his life to bringing these arts to whoever needed them. He did regain his health and he did keep his promise. He spent the rest of his life travelling around the world to lead workshops. Guan Yin's name translates as 'one who hears the sufferings of the world.' She is 'the all-compassionate goddess of mercy' (FLK, 2008, p. 117). Guan Yin is usually shown in figurines and paintings holding a vase in which she collects tears from those suffering (see cover photograph of the present book). Sometimes she is depicted holding the vase vertically in which she has collected tears; at other times, she is shown tipping the vase from which tears flow out. Guan Yin does both.

In the 'three religions' 'high shrines' with three deities depicted (such as the one shown on the cover of *A Path of Dual Cultivation*), one Buddhist, one Confucian and one Taoist deity are presented. Guan Yin is typically the Buddhist deity, while the Jade Emperor is the Confucian deity and Immortal Lü the Taoist deity. Guan Yin 'represents compassion' (FLK, 2008, p. 131). The Jade Emperor is 'the supreme ruler of Heaven' (FLK, 2008, p. 123). He represents 'the quality of virtue' (FLK, 2008, p. 131) and is 'considered the personification of the virtues of loyalty, devotion and propriety' (FLK, 2008, p. 114). Immortal Lü represents wisdom and is often depicted holding a fly whisk with which he brushes away illusion. He is 'the synthesizer of the three philosophical and religious systems of China: Taoism, Buddhism and Confucianism' (FLK, 2008, p. 132; see also pp. vi, 11 and 125–130).

The first chapter of *A Path of Dual Cultivation* concludes that 'these three great traditions embody teachings of practical living; how we should interact with others, with ourselves, and nature' (FLK, 2008, p. 16). Taoist teachings about interactions with nature gave rise to Taoist ecology (see Giblett, 2008a, pp. 178–185). Taoist ecology is founded on the Chinese philosophy of five 'elements,' or phases, or 'real forces' (FLK, 2008, p. 133). Western culture is founded on the Greek philosophy of four elements (see Giblett, 1996, pp. 156–162). The difference between them, and the elements included in and excluded from them, has had profound implications for both cultures' interactions with nature. The Chinese five elements are wood, fire, earth, metal and water. The Western four elements are earth, air, fire and water. Chinese philosophy excludes air, perhaps because it is all-pervasive, whereas for Western philosophy it represents the endpoint of sublimation. Both cultures value sublimation, though for Taoism sublimation involves transformation of the body back to the original body in the womb, whereas Western sublimation involves transformation of the body into the mind and the sublime heights of art and theory. Western philosophy excludes wood and metal, perhaps because they are just raw materials to be extracted and cut up, the stuff of nature, whereas for Chinese philosophy they are repositories and channels for the circulation of energy in the body of the earth.

Qi, or internal energy, is a central concept in this regard. *A Path of Dual Cultivation* defines *qi* as 'the vital energy or life force [. . .], the energy that pervades all things in the cosmos' (FLK, 2008, p. 16, n.1). In Taoist terms, there is no dead or lifeless matter distinct from living matter; all matter is living. There are no non-living beings or things for living beings to exploit as non-living raw materials. *Taijiquan* and other internal Taoist arts of health, such as chanting and meditation, involve the production and circulation of *qi*. *A Path of Dual Cultivation* says that 'in Taoist meditation the *qi* is accumulated, cultivated, transformed and channelled through the body to massage the internal organs. [. . .] Scripture chanting also plays an important role in aligning the body, regulating the breath and massaging the internal organs' (FLK, 2008, pp. 7–8). Taijiquan complements chanting as the 'emphasis on alignment, [and] strengthening and straightening the spine' in taijiquan 'helps promote proper posture and stamina for chanting' (FLK, 2008, p. 8).

If any text exemplifies the polytheistic nature of Taoism in general, and of FLK, and the Society in particular, it is the *Ten Thousand Buddhas Sutra* (FLK, 2012). You can't get much more polytheistic than 10,000 Buddhas! The back-cover blurb of the book states that *Ten Thousand Buddhas Sutra* is 'the principal text of the Three Religions traditions from which Fung Loy Kok developed. As a sacred scripture, it is meant to be chanted together with others in ceremonies that are a form of dual cultivation of body and mind' (FLK, 2012). Chanting complements other Taoist arts of health, such as *taijiquan* and meditation (FLK, 2008, pp. 7, 8 and 108).

Taoist Tai Chi™ *taijiquan* and meditation go hand in hand. Master Moy said that 'if you want your taijiquan to advance you must practise Taoist mediation' (FLK, 2008, p. 6). *Taoist Tai Chi*™ *taijiquan* and meditation are complementary arts of health – both physical and mental (see FLK, 2008, p. 7). Tai chi is often called 'meditation in motion' and 'cultivating stillness in motion,' whereas mediation involves stillness and 'cultivating motion in stillness.' *A Path of Dual Cultivation* states that 'one of the goals of Taoist meditation is to achieve a state of stillness and centredness by emptying all thoughts from the mind' (FLK, 2008, p. 6). Achieving this state involves 'correction of posture' (FLK, 2008, p. 7) in the *Taoist Tai Chi*™ internal arts of health and dual cultivation, including *taijiquan* and meditation. *A Path of Dual Cultivation* goes on to state that 'diligent practise of these internal arts helps to align the spine, improve flexibility in the hips and pelvis, and to stretch, tonify and relax the tendons and connective tissue' (FLK, 2008, p. 7). When 'the arms, legs and spine are connected [. . .] movement of the spine propels movement of the arms and legs' (FLK, 2008, p. 10). The spine is 'the engine of movement.'

The teachings of FLK, and the Society focus especially on the arts of health, such as *Taoist Tai Chi™ taijiquan* and meditation. Taoism is religion in action and contemplation. A brief biography of Master Moy Lin-shin (1931–1998) near the beginning of *A Path of Dual Cultivation* states that 'Master Moy was careful to point out these teachings were more than just a set of movements; they encompassed practical wisdom for living, drawing upon the richness of Chinese culture and the Taoist tradition' (FLK, 2008, p. ix). Conversely, the teachings of the three Chinese schools of Taoism, Buddhism and Confucianism as expressed in FLK and the Society are not just doctrines to study, or even ethical guidelines to follow, but teach a set of movements encompassing embodied practical wisdom and giving health benefits. The title of *A Path of Dual Cultivation* points to the integration of body and mind/spirit in these teachings and to cultivating both by following a path. In them there is no body and mind/spirit dualism as in the modern West (see Giblett, 2008a, chapters 1, 2 and 10).

The path of dual cultivation that FLK and the Society lay out involves 'internal alchemy' defined as 'a process of transformation which changes the body and mind into higher levels of functioning. In the Taoist tradition of internal alchemy, mind and body are not split into separate independent units as in traditional Western thought' (FLK, 2008, p. 4). Arguably in traditional Western thought mind and body were not split into separate independent units, but were integrated. They were split in and by modern Western thought first enunciated by the seventeenth-century French philosopher Rene Descartes (Giblett, 2008a, pp. 23–24). The late fifteenth-/early sixteenth-century Italian artist, anatomist and inventor Leonardo da Vinci was on the cusp of this split (Giblett, 2008a, pp. 20–23). In pre-modern (or traditional) Western thought and culture, body and mind (and land and water) were integrated (Giblett, 2008a, chapters 1 and 2). Body and mind, land and water are still integrated in traditional, pre-modern non-Western thought and culture, such as Australian Aboriginal cultures (Giblett, 2008a, chapter 11, 2009, chapter 1, 2011, chapter 11). In the Taoist tradition of internal alchemy, mind and body are integrated and participate in the tradition of other cultures, as well as counter modern Western dualism and provide an alternative to it.

Modern Western dualism and medicine regard the body as a machine. When the machine breaks down, it is repaired and parts are replaced. In traditional Chinese (Taoist) medicine the body is a system of energy flows; disease is blockage, and treatment is unblocking (see Giblett, 2008a, pp. 158–161). The body in Taoist internal alchemy is what *A Path of Dual Cultivation* calls 'the vehicle for the Return to the Tao' (or Source; FLK, 2008, p. 52) in mutuality with nature, whereas in modern Western thought the body is the vehicle and vector for moving into and colonizing the future with the drive for mastery over nature. The modern Western body as machine achieves its apotheosis in the Fascist body, the hard, metallic body of rugged surface. It is the vehicle and vector by which modern Western culture colonizes the body and the future (see Giblett, 2008a, chapter 6). By contrast, the Taoist body of soft and supple depths and surfaces returns the body and mind to the body of the past as an individual and as a species decolonizing both the body and the future. Whereas in the modern West theory and practice are separated, 'the theoretical and the practical aspects of Taoism are skilfully woven together' (FLK, 2008, p. 84).

One way in which the polytheistic Taoism guides interactions with nature within and without us, and brings them into greater affinity with nature, is through chanting and observing Laozi's 'Sutra of Inspiration,' 'considered to be one of the fundamental texts of Taoist scripture' (FLK, 2008, p. 87). The Sutra involves 'accumulating good works' and repudiating 'evil deeds' (FLK, 2008, pp. 89 and 90). Some of these have definite environmental overtones in line with current thinking about plants and animals, such as 'do not hurt trees, grass, and insects. Be empathetic to the suffering of others. Delight in the joys of others' (FLK, 2008, p. 89). Posters around Melbourne, Australia, mounted by animal liberationists are based on the same idea of non-human animals as sentient beings and invite the same response of empathy with their suffering. Evil deeds in this Sutra include 'shooting animals that fly and run. Frightening worms and animals that crawl. Filling in burrows and turning over nests. Injuring young animals and breaking eggs' (FLK, 2008, p. 91). This view of animals contrasts with the biblical interdiction that 'every swarming creature is an abomination' (*Leviticus* 11: 41). Other 'evil deeds' include 'burying insects. Hating. Using chemicals to kill trees' and 'swearing at the wind and rain' (FLK, 2008, p. 93). These deeds are evil as they go 'against the harmony and flow of things' (FLK, 2008, p. 93), against, in other words, the Tao.

A powerful way in which the polytheistic FLK and the Society also guides interactions with nature within and without us, and brings them into greater affinity with nature, is through what is called 'the Five and Eight Animals training.' An entire chapter of *A Path of Dual Cultivation* is devoted to 'the Five Animal Forms' and 'the Eight Animal Appearances' (FLK, 2008, chapter 5). *A Path of Dual Cultivation* explains that 'the Five Animals are referred to as Forms because they represent the more abstract aspects of taijiquan, such as bone strengthening, tendon transformation, spinal movement and accumulation of *qi*' (FLK, 2008, p. 69). These aspects are more abstract in the sense of being 'more internal in nature' as they 'represent the transformations that occur inside our bodies as a result of our taijiquan training' (p. 69). They train 'the internal structure of the body' (p. 69) not visible to the naked or untrained eye. 'The Eight Animals are referred to as Appearances because they describe the more concrete aspects of taijiquan such as abdominal relaxation, sitting low [opening the pelvis and sinking the tailbone], balance, shifting weight, and circulation of blood to the extreme parts of the body' (p. 69). These aspects are more concrete in the sense of being 'more external in nature' as they 'describe the body postures and appearances that result from practising taijiquan' (p. 69). They are 'behavioural appearances' (p. 69) visible to the naked and untrained eye. All the animals can be experienced by the diligent practitioner.

The Five Animals of the forms are: tiger, leopard, dragon, snake and crane. The Eight Animals of the appearances are: bull, deer, horse, lion, monkey, ape, elephant and bear (FLK, 2008, p. 69). *A Path of Dual Cultivation* goes on to elaborate in more detail about the Five Animal Forms:

1 'The tiger represents power and strength. The strength of the tiger is in its bones [. . .] the centre of gravity of its body is low. [. . .] The training of the tiger is reflected in open hips, enlarged pelvic bone, strength in the legs, and increase of marrow and calcium in the bones' (FLK, 2008, pp. 70–71);
2 'The leopard is an animal whose speed lies in its ability to spring and leap.' To do so, 'the tendons mush very relaxed, flexible and elastic. [. . .] The leopard thus represents tendon training' (FLK, 2008, pp. 71–72);
3 'The dragon is an animal that has a very flexible spinal column. In fact, its entire body is the spinal column.' In the dragon dance, the dragon curls and unwinds, contracts and expands. 'The training of the dragon is the training of elasticity of the spinal column' involving 'movement of every vertebrae' (FLK, 2008, p. 72; see also pp. 73–74);
4 'The snake [. . .] uses the turning of the spine' to move. 'The snake also represents the spinal training.' Together with the expansion and contraction of the dragon, the turning of the snake produces 'spiral motion with upward and downward movement combined with rotation' (FLK, 2008, pp. 74–75);
5 'The crane is an animal with exquisite poise and balance. It is also the symbol of longevity in oriental cultures. In taijiquan, the crane represents the accumulation of *qi* and sense of balance' (FLK, 2008, p. 75).

A Path of Dual Cultivation also goes on to elaborate in more detail about the Eight Animal Appearances and how they relate to and complement the Five Animal Forms:

1 'The bull has a very hard layer of bone on top of its skull and around the crown of its [. . .] In taijiquan training we are trying to accumulate *qi* on top of the head so that it resembles the thick layer of protection on the bull's head.' This training represents the channelling of *qi* and blood upwards to the top of the head. It is 'most closely related to the training of the dragon' (FLK, 2008, p. 76);
2 'The deer represents the channelling of *qi* and blood [downwards] to the base of the spine. [. . .] The training of the deer is closely related to the training of the dragon and the snake' (FLK, 2008, pp. 76–77);
3 'The horse is an animal capable of running long distances without tiring. The secret of its stamina is its relaxed stomach. [. . .] In taijiquan we are training to relax our stomach and intestines so that the internal organs drop lower in the abdomen. This lowering of the organs ensures that the centre of gravity is low and that the mass is located near the base of the spinal column.' This makes for better alignment of the spine with better anatomical posture of the skeletal structure and more efficient physiology of the organs. 'The training of the horse is [. . .] closely related to the training of the snake' (FLK, 2008, p. 77);

4 'The elephant walks in a slow, relaxed way, swaying around gently. The walk also has the appearance of mildly bouncing up and down. This relaxed walk is a sign of elastic tendons. [. . .] In taijiquan training we are training to relax our tendons. [. . .] The training of the elephant is most closely related to the training of the leopard since training in both animals involves "changing of the tendons"' (FLK, 2008, pp. 77–78), an important concept and process in this training (FLK, 2008, pp. 52 and 80);

5 'The ape uses its long arms to swing from tree to tree. It is a master at shifting weight and using it to propel movement. The ape represents training in weight shifting and arm strength. "Strength" does not refer to tension of muscle but to relaxed muscles and tendons. [. . .] Training of the ape emphasizes coordination between the arms and legs and the rest of the body [. . .] the limbs must be connected to the spine [. . .] because the locus of weight-shifting is at the base of the spine. [. . .] The training of the ape is closely related to the training of the leopard' (FLK, 2008, p. 78);

6 'The secret of the monkey's agility lies in its legs. [. . .] The monkey represents strength in the legs. [. . .] The monkey is a very acrobatic animal. [. . .] Acrobatic movements are founded on an enhanced sense of balance, knowing how to balance different parts of the body in stillness and movement to achieve equilibrium. In taijiquan, this kind of balance is founded on relaxed tendons, open joints, and low centre. The training of the monkey is related most closely with the training of the leopard' (FLK, 2008, pp. 78–79);

7 'Bears adopt a low posture when they walk, as if they are sunk into their hips. This low centre of gravity gives the bear a stable stance and strength in its legs and abdomen. [. . .] The training of the bear is the training of [. . .] dropping and springiness in the hips. It requires open hips and a strong pelvic structure. Therefore, the training of the bear is closely related to the training of the tiger' (FLK, 2008, pp. 78–79);

8 'The lion is known for its roar which resonates over long distances. The "lion's roar" in Taoist internal alchemy and taijiquan training refers to the development and training of *qi*. [. . .] The *qi* is channelled from the navel, brought up through the diaphragm and the throat and radiates with breath and speech. Chanting is a way of training the lion. [. . .] In its emphasis on the accumulation and movement of *qi*, the training of the lion is closely related to the training of the crane' (FLK, 2008, p. 79).

A Path of Dual Cultivation concludes its account of the animals by stating that 'becoming the animals is a big step in the path of Return' to 'our Source of Being. [. . .] The Tao is our "Source of Being" [. . .] To return to the Tao, we must reconnect ourselves with the Source' (FLK, 2008, pp. 80 and 83; see also pp. 50–51). Reconnecting with the source ourselves means transforming ourselves – body and mind – into our original nature through the internal alchemy of Taoist training and by following the path of 'dual cultivation of body and mind [which] is unique to Taoism' (FLK, 2008, p. 106). Other schools of teaching emphasize cultivation of either body or mind, whereas Taoism integrates them and creates a path of dual cultivation of both working together and complementing each other. Taijiquan training can return us to our original nature and enable us to 'regain the health of the Original Body' in which 'our skin and bones and body resemble those of a baby, flexible, supple and spontaneous' (FLK, 2008, pp. 10 and 15). Taoist internal alchemy 'starts with transformation of the physical body' (FLK, 2008, p. 107), and not with the mind.

The animal forms are the latent, deep structures of internal processes within the body, whereas the animal appearances are manifestations in external actions visible on the surface of the body. These features point to another distinction between the modern Western body and the traditional Taoist body beside mind/body dualism or holism. The modern Western body is a surface of inscription for the law, pain and pleasure to access and train the mind, whereas the traditional Taoist body integrates surface and depth and trains them together. The modern Western body is trained in exercise in which the mind directs the body, whereas the traditional Taoist body is trained in arts of health in which the mind and body are integrated and function together as a whole. The modern Western body participates in sport in which the body is imprisoned in demarcated spaces and in mechanically measured time, whereas the traditional Taoist body plays freely in open, fluid space and flowing, rhythmic time. The modern Western body strives for mastery over itself and the world outside itself, whereas the traditional Taoist body desires mutuality with itself and the world around it. The modern Western body is a machine, whereas the traditional Taoist body is land. The dragon is not only an animal form in the Taoist arts of health representing the spine in the Taoist body, but also the backbone of the land in the Taoist arts of Feng Shui and in Taoist ecology. Feng Shui is the Taoist art of healthy land, mind and body working together. The Taoist 'picture of the internals' of the embryonic human body is also a picture of the body of the earth.[1]

Taoist Tai Chi™ taijiquan has a variety of specific and demonstrable benefits. It is described in an old brochure published by the Taoist Tai Society of Australia as

> a gentle art of health for people of all ages and health conditions. The slow, graceful movements of Taoist Tai Chi increase strength and flexibility and improve balance and circulation. The Taoist style of Tai Chi emphasises stretching and turning in each of the movements in order to gain these and other benefits more effectively.

The brochure goes on to relate:

> Regular practice of Taoist Tai Chi can bring a wide range of health benefits to the muscular, skeletal and circulatory systems. The flowing movements of Taoist Tai Chi serve as a moving meditation that reduces stress and provides a way to cultivate body and mind. Specific health benefits include:
>
> - toning of muscles, tendons and other soft tissues;
> - rotation of the joints through a full range of motion;

- stretching and alignment of the spine to make it strong and supple;
- gentle massage of the internal organs to improve their functioning.

From my experience of these benefits in myself, from reading testimonials written by practitioners attesting to them in newsletters published by various branches of the Society and from hearing anecdotes from practitioners about them too, I thought that there was a wealth of material about the benefits of practising the *Taoist Tai Chi*™ internal arts of health that was unknown to the outside world and that could be explored and showcased to it. In 2003 I travelled to North America and conducted interviews with practitioners about the benefits that they had gained from practising the *Taoist Tai Chi*™ internal arts. These interviews resulted in the book, *Health Recovery: The* Taoist Tai Chi™ *Way* (Giblett, 2008b). The book is dedicated to the memory of Master Moy. The frontispiece of the book is a photo of the columbarium at the International Taoist Tai Chi Centre north of Toronto in Canada in which the ashes of Master Moy and others are housed. Further information about Master Moy, the Fung Loy Kok Institute of Taoism/Taoist Tai Chi Society, classes and other activities is available on its website www.taoist.org. New members are welcome.

Note

1 All these aspects mentioned in this paragraph receive a systematic, often chapter-by-chapter discussion in Giblett (2008a). See also Giblett (2009, pp. 151–152).

References

Fung Loy Kok Institute of Taoism. 2008. *A Path of Dual Cultivation: Teachings of the Fung Loy Kok Institute of Taoism*. Toronto: Fung Loy Kok Institute of Taoism.

Fung Loy Kok Institute of Taoism. 2012. *Ten Thousand Buddhas Sutra*. Fung Loy Kok Institute of Taoism, trans. Toronto: Fung Loy Kok Institute of Taoism.

Giblett, R. 1996. *Postmodern Wetlands: Culture History Ecology*. Edinburgh: Edinburgh University Press.

Giblett, R. 2008a. *The Body of Nature and Culture*. Basingstoke: Palgrave Macmillan.

Giblett, R. 2008b. *Health Recovery: The Taoist Tai Chi*™ *Way*. London: Shepheard Walwyn.

Giblett, R. 2009. *Landscapes of Culture and Nature*. Basingstoke: Palgrave Macmillan.

Giblett, R. 2011. *People and Places of Nature and Culture*. Bristol: Intellect Books.

Index